The
Noise & Acoustics Monitoring
Handbook

First Edition

Coxmoor Publishing Company's

Machine & Systems
Condition Monitoring Series

The
Noise & Acoustics Monitoring
Handbook

First Edition

Author: Robert J. Peters

Series Editor: Trevor M Hunt

ISBN 1 901892 04 2

Other books in the Machine & Systems Condition Monitoring Series include:

Vibration
Wear Debris Analysis
Thermography
Corrosion
Appearance & Odour
Oil Analysis
Acoustic Emission & Ultrasonics
Level, Leakage & Flow
Load, Force, Strain & Pressure
Power, Performance, Efficiency & Speed
Temperature
Condition Monitoring – An Introduction & Dictionary

Published by
Coxmoor Publishing Company
PO Box 72, Chipping Norton,
Oxford OX7 6UP, UK
Tel: +44 (0) 1451 - 830261
Fax: +44 (0) 1451 - 870661
E-mail: mail@coxmoor.com

Printed in Great Britain by Information Press, Oxford, UK

Acknowledgements

The editor and author would like to acknowledge the help given to them in preparing this book from the following sources:

Peter Wilson - Industrial Noise & Vibration Centre (INVC)
Gunnar Rasmussen - Danish State Railways
John Shelton - AcSoft Ltd
Steve Worley - Lucas Industrial Noise Centre, Lucas Diesel Systems
Alan Watling - British Energy

 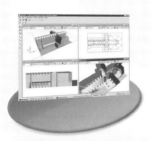

Preface

The Coxmoor *Machine & System Condition Monitoring Series* is a library of books introducing the many disciplines of condition monitoring. However, it goes much further than basics; each volume enables the reader to sense the practical means of monitoring in real situations.

Condition monitoring is not new. Indeed for centuries, operators of vehicles must have detected a deterioration of their equipment before the inevitable final failure. But what is new is the much larger choice open to the user and the considerable advance in all the techniques in recent years.

New ideas, of course, have become necessary because of the tighter levels in manufacture and build, with machines running at higher speeds and greater stresses. In vibration, for instance, damping used to be such an inherent component of build that those waves which were generated were so quickly absorbed that nothing was conveyed. At the other extreme, sound generated from faults in many machines was totally saturated by the background noise from poor fitting systems and components!

Today, machine efficiency is high, and this means the greater is the loss when there is a break-down. Fortunately, monitoring no longer needs to be the monitoring of failure, or even the monitoring of something which eventually will fail, but rather the monitoring of machinery and systems so that they do not fail! Or, if they do, they will fail at the most convenient time, allowing cost effective replacement.

Condition monitoring may thus been seen as providing information which can be utilised in several ways -

1. The identification of the state of the machinery which is of immediate interest and relevance to the maintenance personnel in scheduling inspection and repair of equipment.
2. The planning of future operations by senior management, or for establishing the cost effectiveness of the overall maintenance activity.

3. A feedback on the performance of items manufactured by original
   equipment manufacturers (OEMs) or sold by major component suppliers.

Each volume in the series is self-contained and adheres to a consistent style
enabling ease of reference between the volumes. The complete set will be invaluable to students, maintenance & monitoring personnel, and to managers needing
to know what is available and how to put it into practice.

The following is the list of volumes in this Series:

Vibration
Wear Debris Analysis
Thermography
Corrosion
Acoustic Emission & Ultrasonics
Level, Leakage and Flow
Noise & Acoustics
Load, Force, Strain & Pressure
Temperature
Oil Analysis
Power, Performance, Efficiency & Speed
Condition Monitoring - Introduction & Dictionary

Some readers may only need to read one of the volumes - because of its
immediate relevance to the task in hand; other readers may need to examine more
in order to find the technique best suited to their machine or system. The inclusion of a chapter on the applicability of the technique and case studies, will provide the necessary lead to the best choice.

The majority of each volume is devoted to the actual technique, covering the
basic concepts with some theory (but not too much!), the means by which the
theory is put into practice with descriptions of real instrumentation. Although
there is no suggestion of a 'Which?' guide, advantages and disadvantages are also
discussed, and the extent of, or limit to, their use

The authors are individual experts in the subject, but in the process of providing the text consultation with instrument suppliers and industrial users has
occurred; a full listing of these is included to enable the reader to get further
information from suppliers. In order to avoid excessive bias to any one manufacturer, the authors, as far as possible, have been selected from consultants and
teaching establishments.

In approaching the subject of condition monitoring, it is important to realise
that it is the unseen and unwanted component in life cycle costs. So often it is
rated as the uneconomical aspect of operation which is the last to be considered.

This should not be! It is easy to get a machine going, but to keep it going over a long life, effectively and efficiently, can only be achieved with the help of condition monitoring. Condition monitoring is the key to reliability.

To look forward in faith may be difficult. But for those who have used monitoring extensively and sensibly, there is that privilege of being ultimately rewarded. Condition monitoring makes sense.

Trevor Hunt (Series Editor)
November 2001

Automatic Calibration of Sound Pressure Measurement Equipment providing:-

- *Microphone Open Circuit Sensitivity*

- *Microphone Frequency Response*

- *Magnitude and Phase Response*

- *Insert Voltage Calibration Technique*

- *Electrostatic Actuator Testing*

- *Free Field Correction Curves*

- *User-created Calibration Certificate*

- *Results Stored into ODBC Database*

Calibration in approximately 3 minutes!

Contents

Notation and Symbols
used within the book

[See also the next table for descriptions of multi-notations]

Eng.	Meaning	Gk.	Name	Meaning
A	Weighting	α	Alpha	Absorption coef.
B		β	Beta	
C,c	Weighting (C), Speed of sound (c)	Γ,γ	Gamma	
D,d	Diameter	Δ, δ	Delta	Small change
E,e	Mod. of Elasticity, 2,71828	ε	Epsilon	
F,f	Fast weighting (F), Frequency (f)	ζ	Zeta	
G		η	Eta	Efficiency
H,h	Height, Thickness	Θ	Theta	
I,i	Impulse (I), Electrical current	ι	Iota	
J,j	$\sqrt{-1}$	κ	Kappa	
K,k	Stiffness, Kelvin, kilo	Λ, λ	Lambda	Wavelength
L,l	Level	μ	Mu	Micro (10^{-6})
M,m	Mass, Mega			
N,n	Percentage (N), Rotational speed (n)	ν	Nu	
		ξ	Xi	
O		o	Omicron	
P,p	Peak (P), Pressure (p)	Π,π	Pi	3,14159
Q,q		ρ	Rho	
R,r	Room constant (R), Radius (r)	Σ, σ	Sigma	
S,s	Slow weighting (S), Surface area	τ	Tau	
T,t	Period, Time, Temperature	υ	Upsilon	
U,u	Velocity (u)	Φ,ϕ	Phi	Magnetic flux
V,v	Volume, Voltage, Velocity	χ	Chi	
W	Power	Ψ,ψ	Psi	
X,x	Displacement	Ω,ω	Omega	Angular velocity
Y,y	Displacement			
Z,z	Displacement			

Multi Notations

Type	Notation	Description	Comments
Room	R_c	Room constant	See Eq. 2.11
Sound pressure	L_p p_o L_A	Sound pressure level (SPL) Reference value A-weighted SPL	
Sound power	L_w W_o	Sound power level Reference value	
Sound intensity	L_I I_o	Sound intensity level Reference value	
Sound energy	L_{AE} SEL	A-weighted sound energy Sound exposure level	
Scale	dB dBA dB(LIN)	Decibel scale A-weighted measured dB Linear weighting	Sometimes dB(A)
Time weightings	F S	Fast time constant Slow time constant	
Noise indices	$L_{Aeq, T}$	Continuous equivalent noise level	Over time T (which is stated, e.g. 5min). *see section 2.4.5*
	$L_{AN, T}$	Percentile level	Time exceeded for N%
	L_{Amax} L_{AE}	Maximum sound level Single event noise level	Also called SEL

CHAPTER ONE

INTRODUCTION
TO
NOISE & ACOUSTICS
MONITORING

1. Introduction to Noise & Acoustic Monitoring

1.1 INTRODUCTION

Mankind has been used to auditory clues in order to be alert to changes in the environment since the beginning of the human race. We have all noticed when a piece of machinery does not sound right, whether it is the car, the vacuum cleaner or the central heating boiler. Depending on our inclination, curiosity or determination, we may have attempted to diagnose the cause of the problem using our ears to test how the noise changes for various operating conditions. For example, in the case of a noise in a car we may have investigated whether the noise disappears, gets better or gets worse if we turn a corner, change gear, accelerate, etc.

Many experienced machine operators in the past have used their ears not only to detect faults in their equipment, but also to optimise, or to tune its performance.

Of all the different techniques available for monitoring machinery and systems it is probable that noise commenced first. Our sense of hearing can detect faults which cannot be seen.

Given that much fault diagnosis and condition monitoring has developed in the past from the use of our ears it is perhaps surprising that audio-acoustic signals have not progressed further in formal condition monitoring systems these days.

There are many reasons for this. Perhaps the most fundamental is that much machinery noise originates from a source of vibration caused by the operating forces in the machine. Direct measurement of vibration, rather than the noise that it produces, usually gives a signal that is more suitable for condition monitoring purposes than that from a microphone. Vibration, which is the subject of a separate book in this series, is of one of the most widely used methods of condition monitoring.

An interesting attempt to make the most of vibration and noise monitoring occurred in the 1950s with an engineer's stethoscope called the Bin-Aural (see **Figure 1.1**); instead of having a single probe it had two, one for vibration with a solid contact pointer (called the 'Tectoscope'), and the other for sound with a non-touching conical mouthpiece (called the 'Tectophone'). Although considerable claims were made for the Bin-Aural - like the detection of gear or bearing faults - it depended very much on the ability of the operator to remember what he had heard previously when the machine was operating correctly. Perhaps those were still the days of the 'skilled operator'. It should be noted, however, that the 'skilled operator' still exists medically where the use of a stethoscope can identify the early signs of human body failures, such as, heart valves (a swishing noise), lungs (wheezing or gurgling noise), etc.

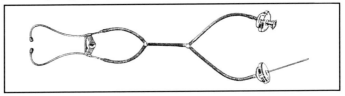

Figure 1.1 - The Bin-Aural [The Capac Company Ltd]

In the 1960s considerable success was claimed in the USA at the Boeing Company for the detection of faults in aero-engines by acoustic analysers. This was followed in the UK by similar work in Rolls-Royce and Bristol Siddeley Engines on Spey, Dart and Olympus engines. This engine research had the objective of being able to monitor such parts as bearings, blades, shafts and gears from one or more carefully placed microphones; the engine frequencies being recorded at idle speed. Although the best analysers were used at the time, with bandwidths down to 0,2 Hz and covering a range of 9 Hz to 5500 Hz, no significant differences were detected between a 'good' engine and one with artificial faults included. (The primary problem was that the difference between any two 'good' engines was greater than the difference caused by the fault!). The conclusion at the time [**Mucklow (1970)**] was that "acoustic diagnosis is not easily achieved when applied to complex rotating machinery".

Since 1970 not only have techniques for analysis greatly improved but our awareness of the relevance of each type of monitoring has grown significantly. The *Coxmoor Series* of books covers well over a score of quite different methods and, whilst there is some overlap, each has its own applicability. Noise is one such technique. Yes, it may have been considered the be-all-and-end-all of techniques - just like all the others - even with its limitations like the others. But it has its advantages!

1.2 THE PROCESSING AND ANALYSIS OF NOISE SIGNALS

1.2.1 Human listening

Our human hearing system is capable of detecting very small changes in our aural environment - much smaller than can be detected using the most sensitive sound level meter. We can use this ability to detect changes in the condition of machinery.

When listening to the noise from a machine our attention may be attracted by a number of possible types of change, including, for example:

- The introduction of a new sort of noise, not previously present, e.g., a new type of hum, whine, whistle, or hissing, knocking or tapping sound.
- A change in the volume or loudness of the existing noise which might indicate to us that the source of noise is either moving closer or further away, or, over a longer time scale, the gradual deterioration of the bearing of a machine, through wear.
- A change in the character of the noise, for example, in the frequency content, which might indicate a change in machine speed or of load conditions. We notice such changes everyday, for example when driving cars or operating lawnmowers, vacuum cleaners or DIY power tools.
- A change in the temporal pattern of a sequence of intermittent sounds, such as in the drip of a tap, the 'tick-over' of an engine, or the rhythm of a piece of music.

1.2.2 Human/machine interaction

Sometimes, in a variation on simple passive listening techniques, we can learn about our environment by sending out a signal and listening to the response. This is what we do when we tap a wall in order to detect whether a particular area is hollow or solid, or when we clap our hands in order to get a 'feel' for the acoustics of a space. It is also what a musician does when tuning an instrument, and what an engineer does when tapping a steel structure, e.g., a wheel, to test the quality of a welded joint. Similar techniques are sometimes used by the discerning shopper choosing crystal glassware or fine china, or by the foundryman when testing the 'soundness' of a bell casting.

Essentially in these 'excitation/response' types of test we are listening for changes, either in the natural frequency (or combination of natural frequencies) of the test object, or for the rate at which the natural frequencies, i.e., the 'ringing' of the system, dies away (dependant on the amount of damping in the system). Both approaches can give valuable information about condition.

1.2.3 Instrumentation

Many of these different 'listening and detecting' strategies, which we use on a day to day basis, almost without thinking, are the basis of numerous forms of acoustic condition monitoring testing, described in the following chapters.

In a condition monitoring system using acoustic signals the microphone takes the place of our ears, and electronic signal processing acts as a very crude form of brain. Obviously, the powers of even the most sophisticated electronic analysis, processing and pattern recognition systems are vastly inferior to those of our brain, which is why we can notice changes in sounds which are undetectable by any man made device. Nevertheless, as shown in the following chapters, it is possible to detect a wide variety of changes in machine condition, using suitable instrumentation.

Figure 1.2 - An advanced noise detection arrangement [Brüel & Kjær]

Figure 1.2 shows the measuring of noise as a vehicle passes by a test station using an advanced measuring system. In this case the total noise derives from several sources, e.g., engine noise, tyre/road interaction, aerodynamic noise, etc. The analysis of the detected signature can be undertaken in a number of different ways as will be explained later. It should be noted also that a number of other tests on vehicles are described later in Chapter 5, both during the time of production and in service life.

1.2.4 Sounds of failure

Many of the 'acoustic' techniques which are used for condition monitoring utilise the ultrasonic frequency range, and are dealt with in a separate book in this series, which also includes the further frequency range of acoustic emission in the one volume.

This volume is restricted to the commonly accepted frequency range associated with what the human ear can detect, namely from around 20 Hz to about 20 kHz. However, it must be noted that whereas a young child may be able to hear up to almost 20 kHz, the older one gets, the lower the upper frequency becomes; adults in middle age may have an upper limit of only around 16 kHz.

Sound originates from disturbances in an otherwise stationary or uniformly moving solid, liquid or gaseous medium. The disturbances, even macroscopic ones, eventually cause pressure fluctuations in fluid media, e.g., air, water, etc., and these are detected by sensing devices such as our ears or a microphone.

In the power generating industry, for example, listening techniques are used to complement the use of more sophisticated instrumental methods, to such an extent that tape recordings of telltale audible features are used to assist inexperienced personnel. Faults in boilers and reactors which can be detected by a trained ear include various types of impact, damage to boiler tubes and reactor internals, aeroelastic instabilities, gearbox problems and cavitation in pumps.

Machinery and system noise, indicative of its condition, can be generated from a variety of sources. **Table 1.1** presents a selection of possibilities, many of which will be discussed further in this volume. Only a few brief examples are given in basic terms to provide a feel for the range which can be covered.

Source	Some examples	Possible fault indications
Sliding surfaces	Pistons	Lubricant depletion
Impacts	Gear teeth mesh	Tooth damage, wear
High velocities	Valves, boilers	Leakage
Rotation	Fan out-of-balance	Blade loss
Liquid changes	Air/water mixing	Breather damage
Loosening	Couplings	Bolt fracture
Low pressure	Pump cavitation	Inlet blockage
Combustion	Air/fuel jets	Part blockage
Fatigue	Bearings	Track pitting
Dimension change	Nozzles	Blocking or corrosion

Table 1.1 - Some sources of 'condition' noise in machinery and systems

1.2.5 Advantages and disadvantages

Compared to vibration monitoring the measurement of noise from machinery
suffers from two serious disadvantages:

1. It is often difficult to separate out the required acoustic signal from
    other completely different sources of noise
2. Reflections from nearby surfaces, and from other machinery, interfere
    with the signal.

To set against these disadvantages, the use of a microphone, as opposed to an
accelerometer, has a number of advantages:

1. The use of the microphone does not require contact with the vibrating
    surface. This is of particular value when the surface is moving (rotating, for
    example) or is inaccessible in some way.
2. Sometimes it is not possible to identify exactly which surface is the
    best one to monitor in order to identify condition-related changes in
    vibration levels; or it may be that there are just too many such possible
    surfaces. In such cases it can be an advantage to use a microphone
    which will pick up sound radiated from all such surfaces.
3. The production of noise is not always associated with a vibrating
    solid surface. The detection of aerodynamically produced noise from
    jets is one example. The detection of the high frequency noise from a
    leaking pneumatic, hydraulic or process fluid system in a factory is
    more easily accomplished using a microphone (or even a pair of ears)
    rather than an accelerometer.
4. There is a wide range of versatile and accurate instrumentation avail-
    able for the measurement of noise for environmental and occupa-
    tional safety purposes, and much of this may be adapted for use in
    condition monitoring.

Finally, we must realise that even the most sophisticated analyses of
noise signals, which can be performed at present, fall far short of that performed
by the human ear and brain, in distinguishing and recognising subtle changes
in the noise from machinery. It is just that there is a need to have some means
of accurately comparing the problem signal generated today, with what was ac-
ceptably generated yesterday, that instrumentation becomes necessary.

1.3 SCOPE

The aim of this book is to explain the principles and techniques of noise generation and measurement and to indicate the methods available for using noise signals as the basis of a tool for condition monitoring.

Chapter 2 will describe the basic concepts and terminology used in the measurement of noise, and the mechanisms whereby it is generated in machinery.

Chapter 3 discusses the practical difficulties to be encountered when measuring noise. It includes comments on the effect of the acoustic environment on the measurement procedure

Chapter 4 describes the range of noise measuring, analysis, monitoring and recording equipment which is available and suitable for condition monitoring.

Chapter 5 describes actual case studies illustrating the successful application of noise measurement to condition monitoring.

Chapter 6 is a buyer's guide to the manufacturers and suppliers of available equipment and of their products; also included are various services.

Chapter 7 contains a comprehensive glossary of terms, information about relevant standards and a list of references and further bibliography.

It should be understood that the most usual reason for measuring noise, whether in industry, transportation or from a leisure source, is as a first step towards its reduction and control. The terminology of 'technology of noise measurement and analysis' has developed in that use for the control of noise. We shall use the same terminology and technology in this book, although as we shall see in Chapter 5 that the day of the virtual sound level meter has arrived where a microphone signal, with minimal conditioning, can be fed into a lap top computer which can, in principle, be programmed to perform almost any kind of noise analysis that is required.

CHAPTER TWO

BASIC CONCEPTS
AND
THEORY

2. Basic Concepts and Theory

2.1 INTRODUCTION

In this chapter the basic concepts and terminology relating to the nature of sound and how it is described and measured are discussed, in order to introduce the reader to noise measurement parameters which may be used for condition monitoring purposes. The way in which noise is described and measured reflects the fact that in the vast majority of cases the measurement is related in some way to the effect of the noise on people. Thus the condition monitoring engineer may find that the simplest, and cheapest sound level meter which is available may only measure A-weighted decibels, dBA. However, the A weighting, as explained later in this chapter, is designed to simulate human response to noise, and is not therefore necessarily of particular relevance to the measurement of machinery noise for condition monitoring purposes.

Since many of the practical difficulties associated with measuring noise are related to the effects of the measurement environment some further concepts and theory about sound propagation are included in Chapter 3.

This chapter also includes a brief review of the effects of noise on people and society.

2.2 THE NATURE OF SOUND

Sound waves in air consist of minute fluctuations in atmospheric pressure, caused by the vibration of the air particles close to the source of the sound. Because the air has the properties of mass and elasticity the vibration of the air particles, and hence the disturbance in atmospheric pressure is transmitted through the air from source to receiver.

To explain this further it is useful to adopt an acoustician's simple mental picture of air as consisting of a series of interconnecting masses and springs, representing the properties of the inertia and elasticity of the air. Any disturbance to a particular portion of air corresponds to disturbing one of the masses in the simple model, and will cause the mass (i.e., that portion of air) to vibrate. This vibration will be transmitted, via the interconnected springs in the model to the surrounding air, and eventually to the portion of air adjacent to a human eardrum, or a microphone. The vibration of the air particles cause minute fluctuations in air pressure, which when transmitted to the eardrum or the microphone are either heard, or measured, as sound.

2.2.1 The pure tone

The simplest of all sounds is the pure tone such as the sound from a tuning fork, or from the time signal pips on the radio. For such a sound the graph showing the variation of sound pressure with time, called the waveform of the sound, is a sine wave, as shown in **Figure 2.1**. The frequency of the pure tone is the number of sine wave cycles per second, and is measured in Hertz (Hz). The magnitude of the sound pressure fluctuations can be measured, in pascals, as the amplitude of the waveform, which is also, in this simple case, the same as the peak value of the sound pressure.

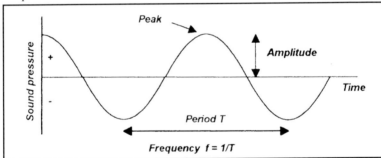

Figure 2.1 - A pure sine wave

2.2.2 Wavelength

In the same way that the frequency of a pure tone characterises its cyclic variation with time the wavelength similarly describes its variation in space. If we consider a pure tone sound passing down a smooth tube or pipe, with no loss of energy due to absorption, then at any moment the variation of sound pressure with position along the tube will also be sinusoidal, as shown in **Figure 2.2**. Two points which are separated by only a small fraction of a wavelength will be almost in phase, and have similar sound pressures at any one moment. As the separation increases so does the phase difference, until when the separation is half a wavelength the sound pressure at the two points will be completely out of phase. The wavelength of the sound is the minimum distance between two points in the pipe for which the sound pressure waveforms will be in phase.

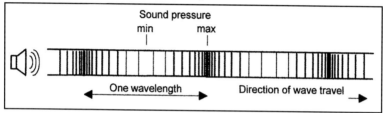

Figure 2.2 - Sound wave propagation

Wavelength λ and frequency f are related by the sound speed, c:

$$c = f\lambda \tag{2.1}$$

For sound in air in the temperature range from 0 °C to 20 °C, c can vary between about 330 m s^{-1} and 340 m s^{-1} depending upon temperature. (In water the speed of sound is considerably higher, typically 1500 m s^{-1}, and even higher in solids.)

Thus, for a fairly low frequency sound, say 100 Hz, $\lambda = 3,3$ m; for a medium frequency sound, say 1 kHz, $\lambda = 0,33$ m, and for a high frequency sound, say 10 kHz, $\lambda = 0,033$ m or 33 mm.

It is the wavelength of a sound that determines how it is scattered and diffracted by obstacles in its path, or how much attenuation is provided by a screen or barrier. In these cases it is the size of the obstacle or barrier compared to the wavelength that is important. An obstacle which is large compared to the wavelength will reflect sound and will cast an acoustic shadow, as is the case with light. Conversely an object which is small compared to the wavelength will almost be 'ignored' by the wave, except perhaps for some scattering of sound, and there will be almost no shadow behind the object. Thus acoustic screens or barriers are only effective for frequencies for which their dimensions are large compared to the wavelength. A consequence is that noise measurements at high frequencies, when wavelengths are smaller compared to the size of the microphone, are much more sensitive to changes in the microphone position or to the angle at which the microphone is oriented towards the source of the sound than is the case when measuring low frequencies.

Another consequence is that it is the size of a noise source compared to the wavelength that determines the directionality, or directivity of the source. When the source is large compared to the wavelength it will radiate sound more or less equally in all directions, but for higher frequencies, when the source becomes, relatively, even larger compared to the wavelength, the pattern of sound radiation becomes more directional. This is why separate small loudspeakers, called 'tweeters' are fitted into high-fi loudspeaker cabinets to radiate the high frequency parts of the sound signal (which would have been 'beamed' in one direction by the larger loudspeaker) more or less evenly in all directions.

2.2.3 Peak and rms sound pressures

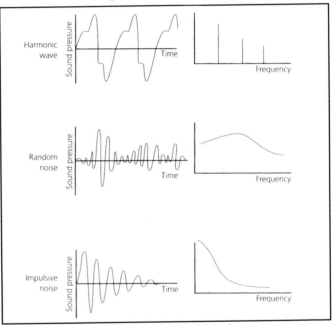

Figure 2.3 - Different waveforms and their frequency spectra

A pure tone contains only one frequency **Figure 2.1**. Most sounds are not pure tones, but have more complex waveforms, such as illustrated in **Figure 2.3**, and may be thought of as a combination of pure tones of different frequencies and amplitudes. It is no longer possible to assign a simple amplitude to such a waveform, and the magnitude of the sound is most commonly expressed in terms of the root mean square (rms) sound pressure. It is also possible to measure the peak or peak to peak sound pressure.

For the pure tone there is a simple relationship between the peak and rms values:

$$rms = Peak / \sqrt{2} \tag{2.2}$$

For more complex waveforms there is no such relationships, but the ratio of peak/rms value, called its 'Crest Factor', is an important characteristic of the sound. Impulsive sounds have a high crest factor. The crest factor of a pure tone is $\sqrt{2} = 1,414$.

2.2.4 The behaviour of sound waves

Sound waves, like all other form of wave, undergo reflection, interference and diffraction.

a. Diffraction

It is the property of diffraction that determines the extent to which sound waves are bent around corners, or around the edge or over the top of a screen or barrier between source and receiver. In these cases it is the size of the diffracting object compared to the wavelength of the sound which is the determining factor.

Diffraction also determines how sound is scattered by obstacles in its path. An example is the way sound is diffracted by a microphone. At low frequencies, where the wavelength of the sound is much greater than its diameter, the microphone responds equally to sound from all directions because the sound waves spread easily around its surface. The microphone becomes more directional at high frequencies when its size is similar to the wavelength of the sound.

b. Reflection and absorption

When the surfaces of the object are large compared to the wavelength, sound waves are reflected, and obey the same laws of reflection as light reflected from a mirror.

Hard surfaces such as concrete, brickwork, plaster, glass and metal (including many machine surfaces) reflect almost all the sound energy incident upon them. In contrast soft fibrous surfaces such as carpets, curtains, upholstered furniture, and slabs of glass or mineral fibre are poor reflectors and absorb most of the sound energy, converting it into heat as a result of frictional processes which occur within the small air spaces between the fibres of the material. The absorption coefficient of a material is the fraction of sound energy which is either reflected or absorbed at the surface of the material. Absorption coefficients range from

1 - for a perfect absorber which will not reflect any sound, to
0 - for a perfect sound reflector, which will not absorb any sound energy.

As well as being frequency dependent the absorption coefficient also varies with the angle of incidence at which the sound strikes the surface, and in the case of fibrous materials, the thickness of the layer. In practice, values range from about 0,01 to 0,05 for hard surfaces at low frequencies, to 0,9 to 0,95 or higher, for thick layers of fibrous material at high frequency.

The absorption of sound by the air is negligible except at high frequencies and for long distances outdoors, and is not usually significant indoors.

c. Refraction

Refraction is the change in direction of a sound wave caused by changes in sound speed, which may occur as a result of changes in temperature or wind speed. Refraction effects can be very important for propagation over considerable distances outdoors, but is not a significant factor indoors.

2.3 HUMAN AWARENESS OF SOUND

2.3.1 The audible frequency range

The audible range of frequencies is from, approximately, 20 Hz to 20 000 Hz (20 kHz). Sound with frequency above 20 kHz is called 'ultrasonic' and frequencies below the audible range are known as 'infrasonic'. The humming noise of a transformer, e.g., from an electricity sub-station, consists mainly of 50 and 100 Hz frequencies. Most fans produce a broad band noise spectrum, but with most of the sound energy in the low frequency range, with the highest levels usually occurring in the 63 Hz, 125 Hz and 250 Hz bands. By contrast many machines produce high frequency noise, with the noise from hand held DIY drills and grinders being highest in the octave bands at 1000 Hz and above. The 'pips' from the time signals on the radio are short bursts of 1000 Hz tones, and the various alarm bleeps signals from computers, and other domestic equipment are at similar frequencies or higher, as are the reversing signal warnings fitted to commercial vehicles.

2.3.2 The audible sound pressure range

The audible range of sound pressures from the threshold of hearing to the threshold of pain is, approximately, from about 0,00002 Pa to 100 Pa. In order to put this range into perspective it should be noted that atmospheric pressure is basically 100 000 Pa (0,1 MPa or 1 bar). Thus it can be seen that sound represents only very small fluctuations in atmospheric pressure. Even at the top end of the audible range, called the threshold of pain, the sound pressure is only one thousandth of an atmosphere. The range of sound pressures, from lowest to highest is enormous, about 5 million to 1. This is a huge measurement range to be covered by just one instrument - the human ear. The difficulties associated with the very large range of sound pressures is one reason for using a decibel scale.

2.3.3 The decibel scale

The decibel scale is a logarithmic scale for comparing ratios of powers, or quantities related to power. In acoustics three separate decibel scales are used, for sound power, sound intensity and sound pressure. These are defined as follows:

Sound power level, $L_W = 10 \log (W/W_O)$ (2.3)
Sound intensity level, $L_I = 10 \log (I/I_O)$ (2.4)
Sound pressure level, $L_P = 20 \log (p/p_O)$ (2.5)

where W_O = reference value of sound power (10^{-12} W)
         I_O = reference value of sound intensity (10^{-12} W m^{-2})
         p_O = reference value of sound pressure (20 x 10^{-6} Pa).

Note that the use of the word 'level' always implies that decibel scale is being used. Thus sound pressure is measured in pascals, but sound pressure level in dB. Similarly, sound power is measured in *watts* but sound power level in *dB*.

Sound power level is a term which describes the noise output from a noise source such as a machine, whereas the sound intensity and pressure levels describe the magnitude of the sound produced at some point in the environment by that machine. This is analogous to specifying the heat generated by an electric fire in kilowatts, but the temperature created in the room by the heat radiated from the fire is measure differently, in degrees.

The factor 20 in one of the above definition arises because in order to represent sound pressure on the decibel scale it is necessary to invoke a fundamental relationship in acoustics, that sound intensity is proportional to the square of sound pressure. The relationship is analogous to that in electrical circuits where electrical power is proportional to the square of voltage, and in vibration, where the energy of a vibration is proportional to the square of the vibration amplitude. The value of p can be usually taken to be the rms value.

Each 3 dB increase represents a doubling of sound energy
Each 6 dB increase represents a fourfold increase
Each 10 dB increase represents a ten-fold increase
Each 20 dB increase represents a hundred fold increase, and so on.

Subjectively a 1 dB increase is only just noticeable, a 3 dB increase is clearly noticeable and a 10 dB increase represents a doubling of loudness. A 20 dB increase would therefore represent a four-fold increase in loudness.

Note that although the various levels defined above refer to absolute decibel scales, fixed by the use of a reference value, it is also possible to use decibels in a relative manner. As an example it would be possible to describe the noise produced by machine A as being 20 dB higher than that from machine B. This description does not specify, in absolute terms the amount of noise produced by either machine, but that one produces a sound power which is one hundred times that of the other.

2.3.4 Background noise effects

A microphone placed close to a machine will respond to background noise at a measurement position as well as to the noise from the machine. In order to understand the significance of the background noise it is necessary to consider how noise levels measured in decibels are combined.

When decibels N different levels: L_1 L_2 L_3 . . . L_N are combined the total noise level may be obtained from the formula:

$$L_T = 10 \log (10^{L1/10} + 10^{L2/10} + \ldots 10^{LN/10}).$$ (2.6)

Alternatively when only two levels are to be combined, **Table 2.1,** which gives a correction factor to be added to the higher of the two levels, may be used.

Difference between the two levels - dB	Correction to be added to higher level - dB
0	3
1	2
2	2
3	1
4	1
5	1
6	1
7	1
8	0,5
9	0,5
10	less than 0,5

Table 2.1 - Combining two noise levels

From **Table 2.1** it can be seen that if one of the two levels is 10 dB or more below the other then the lower level makes a negligible contribution, less than 0,5 dB to the overall level. Background noises are discussed further in Section 3.2.1 a.

2.3.5 Loudness

The loudness of a sound depends on its frequency as well as its sound pressure level, because the sensitivity of human hearing varies with frequency. It is most sensitive in the frequency range from 1 kHz to 4 kHz and least sensitive at lower and high frequencies, as indicated by the Equal Loudness Contours shown in **Figure 2.4**. The A-weighting scale was devised many years ago in an attempt to incorporate the variation in hearing sensitivity into the sound level reading indicated by the sound level meter. It is an electronic filter based on the shape of one of the Equal Loudness Contours. The measured value is called the A-weighted sound pressure level, or the sound level in dBA, and also written as L_A

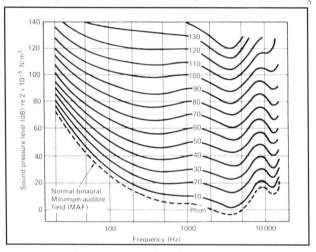

Figure 2.4 - Equal loudness contours

2.4 SOUND ANALYSIS

2.4.1 Frequency weighting networks

Originally three such weighting networks were devised: A, B and C, based on three different contours. Today only the A-weighting is commonly in use, although the C weighting is also used for impulsive noise because it represents an almost flat frequency response. Some sound level meters also allow un-weighted sound pressure levels, i.e., without any frequency weighting, to be measured. This is sometimes called a Linear Weighting, dB(LIN). The A and C Weightings are shown in **Figure 2.5**. The values of A-weighting at low frequencies are high, and thus a sound containing predominantly low frequencies will have a dBA value which is several decibels lower than its dBA or dB(LIN) value, whereas for a high frequency sound the three values will be very similar.

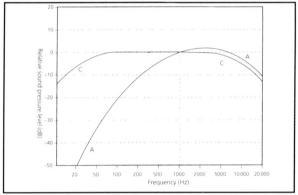

Figure 2.5 - A and C weightings

On a decibel scale the audible range of sound pressures cover a range from 0 to 140 dB. The sound level, in dBA, of some common sounds, illustrating and comparing the audible range, is indicated in **Table 2.2**.

Sound pressure - Pa	Sound level - dB	Typical conditions
100 Pa	134	Threshold of pain
10 Pa	114	Pneumatic drill
1 Pa	94	Noisy street
100 mPa	74	Average speech
10 mPa	54	Business office
1 mPa	34	Library
100 μPa	14	Broadcast studio
20 μPa	0	Reference threshold

Table 2.2 -Pressure and noise levels

2.4.2 Frequency analysis

The frequency content of a noise may be measured using a frequency analyser and the results displayed as a frequency spectrum. The most commonly used method in general acoustical work is octave band analysis which divides the audio-range into a number of contiguous frequency bands, which are equally spaced when plotted on a log frequency scale. Each band is known by its centre frequency. The centre frequency of the bands are:

31,5 Hz, 63 Hz, 125 Hz, 250 Hz, 500 Hz, 1 kHz, 2 kHz, 4 kHz and 8 kHz

Sometimes when greater frequency discrimination is needed one-third octave bands are used: each of the octaves is split into three one third octave bands.

The octave and third octave bands are used to assess or evaluate a noise, or to specify noise control treatments. If a noise contains tonal components and it is required to measure the precise frequencies of the tones, in order to identify the source of the noise for example, then narrow band analysis is needed. Narrow band analysers may operate on a constant percentage bandwidth basis, as with one-twelfth or one-twenty fourth octave bands, or on a constant bandwidth basis as with Fast Fourier Transform (FFT) analysis. An example of a 1/3 rd octave and a narrow band spectrum, in this case a 1/12 th octave, are shown in **Figure 2.6** obtained from a gear box drive. Narrow band analysis of vibration signals from rotating machinery is widely used as a Condition Monitoring tool.

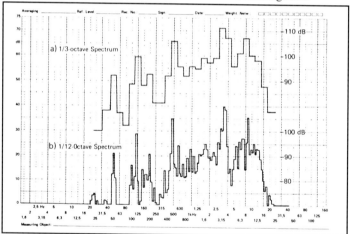

Figure 2.6- Octave and narrow band comparison of the same signal
[Brüel & Kjær]

2.4.3 Level versus time variations

Sometimes the noise level from a machine or from a production process varies with time and the pattern of variation may be used to identify different parts of the process. **Figure 2.7** shows the variation of sound level, in dBA, of the noise from an automatic washing machine over a complete wash cycle at full load; the different parts of the cycle are quite different in noise level and can be identified even at 10 second intervals. **Figure 2.8** shows the corresponding pattern from an industrial guillotine, over a shorter time scale. In both cases particular events within the cycle can be identified. Usually it is the variation of the dBA value with time which is monitored but with modern instrumentation it is possible to track the time variation of any frequency band. If the frequency spectrum of the noise changes with the passage of time, as well as the overall level, it is also possible to display the variation of all three parameters in a 'waterfall' plot.

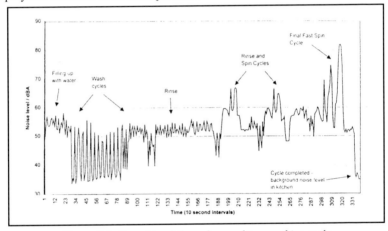

Figure 2.7 - Sound level through a washing machine cycle

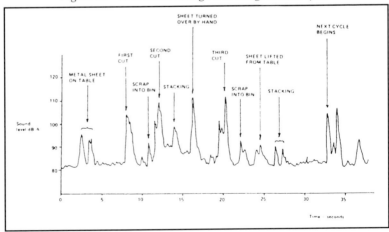

*Figure 2.8 - Sound level with time from a guillotine [**Tomlin (1984)**]*

2.4.4 Fast and slow time weightings

When measuring the instantaneous value of sound pressure level at any one moment two different averaging times (also called time weightings) are commonly used in sound level meters to determine rms levels. A form of running time averaging, called exponential averaging is used. The F time constant is 0,125 s and for S it is 1,0 s.

If the noise is fairly steady with time then the same sound level should be indicated whichever time weighting is used, although any moment to moment fluctuations will be smoothed out more if S is used. If however the noise is transient, such as a motorbike or fast car passing, then a higher level will be obtained using the F weighting than with S.

For transient sounds of even shorter duration such as noise from gunshots, or explosions or an industrial punch-press or a pile-driver there will be even greater discrepancy between the two readings. A third time-weighting, called the Impulse, can also be used for these situations. This has a rise time of 35 ms, i.e., about one thirtieth of a second. Alternatively the peak value may be measured. **Table 2.3** shows the variation between F, I and P readings for a variety of different sounds.

Sound source	Fast - dBA	Impulse hold - dBA (5x)	Peak hold - dBA (5x)
Small car at 60 km/h (inside)	78	79	93
Large car at 80 km/h (inside)	75	76	86
Generating plant 75 hp diesel	100	101	113
Motorway traffic at 15 m	80	81	89
Motorway traffic at 50 m	68	68	76
Train at 70 km/h at 10 m	95	98	106
Train at 70 km/h at 18 m	85	87	94
Aircraft PA 23 cruising (inside)	90	91	100
Aircraft KY 3 cruising (inside)	103	103	112
Pneumatic nailing machine near head	116	120	148
Pneumatic nailing machine at 3 m	112	113	128
Industrial ventilator 5 hp at 1 m	82	83	93
Air compressor room	92	92	104
Large machine shop	81	82	98
40 t punch press near head	93	97	121
Small automatic punch press	100	103	118
Numerically driven high-speed drill	100	103	112
Machine driven saw near head	102	104	113
Pistol 9 mm at 5 m (side)	113	116	146
Shotgun at 5 m (side)	108	111	143

*Table 2.3 - Impulse sound-level meter readings [**Yang & Ellison (1985)**]*

2.4.5 Single figure noise indices

A number of single figure values may be used to describe a time varying period of noise.

The continuous equivalent noise level, $L_{Aeq, T}$ is the constant level of continuous noise which, over the period of time (T) under consideration contains the same amount of A-weighted sound energy as the time varying noise.

The percentile level, $L_{AN, T}$ is the level exceeded for N% of the measurement period. The two most commonly used percentile values are the L_{A10} commonly used as a measure of traffic noise and L_{A90} widely used as a measure of background noise, but some instruments will allow any integer percentile value to be measured. The L_{A1} and L_{A99} are also sometimes used to indicate the highest and lowest levels in a measurement period. The maximum sound level, L_{Amax} is also used. This is the maximum value of the A-weighted sound level occurring within the measurement period. This should not be confused with the peak sound pressure level which is based on the highest value of the waveform sound pressure without application of the rms time averaging or frequency weightings. The, L_{AN} and L_{Amax} are based on rms sound pressure levels and so may vary depending upon whether F or S time weighting is selected. The L_{Aeq} and L_{peak} values are not based on rms values and so do not require time weighting to be specified.

All these values may be measured over almost any period of time ranging from seconds to hours. The length of the time period is often indicated as an additional subscript, e.g., $L_{Aeq, 5min}$ or $L_{A10, 18hr}$.

For a bursts of noise which can be described as 'noise event' the single event noise level, SEL, or L_{AE} is also used. This is the L_{Aeq} value which, over a period of one second would contain the same amount of A-weighted sound energy as the noise event.

2.4.6 Measurement of sound intensity

The direct measurement of sound intensity is based on the relationship between the intensity, I, the sound pressure, p and the acoustic particle velocity, u :

$$I = p\, u \qquad\qquad (2.7)$$

This is analogous to the relationship that electrical power consumed by a device is the product of the voltage across it and the current passing through it (i.e., watts = volts x amps). Since the mid 1980s sound intensity meters have become available which measure sound intensity directly. They operate by simultaneously measuring the sound pressure and acoustic particle velocity at the same point, multiplying the two signals together and averaging over an appropriate time period. The most commonly used type uses a two microphone probe consisting of two precisely matched microphones with a small air gap between their two diaphragms. Two microphones are needed because this enables the sound pressure gradient to be measured, and acoustic theory shows that this is related to the rate of change of acoustic particle velocity with time.

The vector nature of acoustic intensity means that the two-microphone sound intensity probe is directional in its response, giving the maximum signal when the axis of the probe, i.e., the line joining the two microphones, is lined up in the direction of the flow of acoustic energy. In order to measure the total sound intensity passing through a surface, e.g., a window, or a machine panel, the sound intensity readings are averaged over the surface by passing the sound intensity probe over the surface. The meter integrates the sound intensity over the entire surface and indicates whether the net flow of acoustic energy is either into or out of the surface.

It is this ability to identify the direction of flow of acoustic energy, as well as measuring its magnitude, which is the basis of many of the applications of the sound intensity meter.

One of these applications is the measurement of sound power emitted by noise sources. Using the acoustic intensity meter it is possible to measure the sound power level of a noisy machine in situ, in the presence of background noise from nearby machines and also without the need for a specialist acoustic environment, i.e., an anechoic or a reverberant room. Other applications include the location and identification of noise sources and the detection of leaks and weak link areas in the transmission of sound through panels and partitions.

2.5 SOUND PROPAGATION

2.5.1 Sound intensity and the effect of distance

The intensity of a sound, measured at any particular point is the sound power per unit area passing through a surface at right angles to the direction in which the sound is travelling. It is measured in watts per square metre, W m^{-2}.

Sound intensity = sound power / area   [W m^{-2}]   (2.8)

Sound intensity is therefore a vector quantity with both a magnitude and a direction, in contrast to sound pressure, for example, which is defined entirely by its magnitude alone. Sound intensity is, therefore, highly important to the theory of sound propagation.

If the source is assumed to be a simple point source radiating spherical wave-fronts then, at a distance r from the source, the wave-front will have spread over an area of $4\pi r^2$, and the intensity, I, at a distance r from a point source of power W will be:

$$I = W / 4 \pi r^2 \tag{2.9}$$

This relationship incorporates the well known inverse square law which also relates the intensity of other sorts of radiation such as light or radioactivity with distance from the source. Converted into decibels this becomes an equation which is the basis of much noise prediction:

$$L_p = L_w - 20 \log (r) - 11 \tag{2.10}$$

where L_w = the sound power level of the sound source (dB)
L_p = the sound pressure level (dB).

The inverse square law relating intensity to distance becomes, in decibels, a reduction of 6 decibels for each doubling of distance.

This equation is only true for sound from an idealised point source in free field conditions, i.e., in the absence of any reflections from nearby surfaces. The equation may, however, be modified and extended to include the directional properties of real noise sources, the effects of room reflections, and outdoors, the effects of air absorption, ground attenuation, noise barriers and other atmospheric effects on sound propagation. (See also Section 2.4.6 - *Measurement of sound intensity.*)

2.5.2 Real noise sources: near and far fields

Equation (2.10) is based upon a model of the noise source as a simple point source radiating sound energy equally in all directions. Real noise sources have finite dimensions, i.e., they are not points, and they may be directional. However, provided that the measurement position is far enough away from such a source, the variation with distance, in any given direction, will obey the above equation and the 6 dB per doubling of distance rule, see **Figure 2.9**. The region of space surrounding the source where this applies is called the 'far field' of the source.

Closer to the source the contributions from different parts of the source combine to produce a much more complex pattern of sound pressure level variations which is not easily amenable to prediction. This region of space surrounding the source is called the 'near field' of the source.

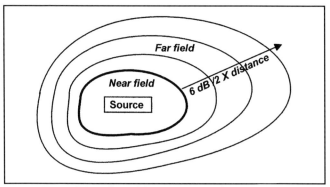

Figure 2.9 - Effect of distance on sound measurement

2.5.3 The behaviour of sound in enclosed spaces

In an enclosed space the sound energy transmitted to the receiver directly from the source is augmented by that arriving after reflection from the room surfaces, and from diffraction by objects in the room. The sum total of all the reflections and diffraction effects is called reverberation. A simple theory of sound distribution in rooms predicts that the sound pressure level due to the reverberation, called the reverberant sound pressure level, is constant throughout the room, and depends on a parameter called the room constant (R_c). R_c depends on the amount of sound absorption in the room:

$$R_c = S \, \alpha \, / \, (1 - \alpha) \qquad (2.11)$$

where R_c = room constant [m^2]
$\quad$ S = the total area of these surfaces [m^2]
$\quad$ α = the average absorption coefficient of the room surfaces, obtained from

$$\alpha = \{\alpha_1 S_1 + \alpha_2 S_2 + ...\} \, / \, \{S_1 + S_2 + ...\}$$

The reverberant sound pressure level is given by:

$$L_p = L_w + 10 \log (4/ R_c) \qquad (2.12)$$

Figure 2.10 shows how the direct and reverberant sound pressure levels, and the total sound pressure level obtained by combining them, varies with the distance of the receiver from the source. It can be seen that close to the source the direct sound pressure level is the dominant component of the total, and decreases with distance at a rate of 6 dB per doubling of distance. At large distances from the source the reverberant component is the dominant one, and the total sound pressure is independent of distance in this region.

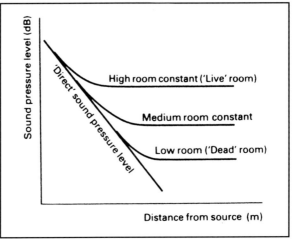

Figure 2.10 - Typical sound pressure pattern for various rooms

Figure 2.10 also shows that changing the room constant can cause the reverberant sound pressure to change, but the sound pressure level close to the source is unaffected

2.6 SUMMARY OF POSSIBLE NOISE 'SIGNATURES' FOR CONDITION MONITORING

It can be seen that a noise may be described and measured in many different ways. Some of the methods which could be suitable for condition monitoring purposes are summarised below.

2.6.1 The acoustic waveform - the graph of sound pressure versus time

This can be obtained by displaying the output from a microphone onto an oscilloscope or similar piece of equipment. Usually the time period of interest is a fraction of a second may be only a few milliseconds. This type of display might be useful to identify the noise generating events and mechanisms occurring within a machine - a series of impacts for example, or the 'firing' sequences in an internal combustion engine, and to monitor changes which may occur in these mechanisms, as a result of wear for example.

The *Coxmoor series* book on vibration, gives examples showing the effects of impacts caused by looseness of components and by electrically generated signals on the vibration waveform of a motor shaft. In principle similar waveforms can be detected by placing a microphone close to the motor, although chapter 3 will indicate some of the practical difficulties in practice.

2.6.2 The sound level versus time history

Bearing in mind that a sound waveform could contain up to 20 000 cycles per second it can be seen that a waveform of more that a fraction of a second may contain too much information to be useful. As explained earlier, the variation of sound level (in decibels) over longer periods of time can also indicate the sequence of noise producing events in a machine and be useful for condition monitoring.

2.6.3 The frequency spectrum - the graph of sound level versus frequency

All forms of rotating machinery produce forces which are repeated regularly, at a rate related to the rotation speed of the machine. This results in a line frequency spectrum which consists of a series of frequencies, at regular frequency intervals, which are related to the harmonics of the shaft rotation frequencies. Narrow band analysis is the form of frequency analysis most suitable for measuring this type of spectrum, and is widely used for vibration condition monitoring, and can be used for noise also. If particular harmonics are shown to correlate best with certain machine conditions (looseness or out of balance, for example) it is possible to specifically monitor the changes in amplitude at these frequencies.

2.6.4 Measurement of noise indices

The simplest type of measurement, but perhaps likely to be the least selective in terms of correlating to any particular machine condition, is of some overall noise measure such as dBA (either instantaneous values or L_{Aeq})

During a noise survey in a factory carried out as part of an assessment of

employee noise exposure the author was able to detect significant differences in
the noise output from some machines in a line of nominally identical pedestal
grinders. It transpired that the noisy machines had badly worn bearings. The measurements were of simple dBA levels, with the microphone held close to each
grinder, at a position where the operators ear would be located.

Often overall noise levels from machinery vary smoothly with change in
some operating parameter such as speed, or load, or pressure, etc., and departures
from a smooth trend as depicted on a graph, can indicate some significant change
in machine condition. As an example the noise level from fuel injection pumps
increases with the pump rotational speed in revolutions per minute, with a graph
of dBA versus speed (plotted on a logarithmic scale) giving a straight line graph.
On one occasion the author was involved in testing a new type of pump which
operated at up to higher speeds than the previous model. The graph of dBA versus
(log of) speed was a straight line, as expected for the usual speed range but de-
parted markedly at the higher speeds making more noise than would have been
predicted from the trend line. It was apparent that a new noise producing mechanism was being brought into operation at the higher speeds. It subsequently turned
out that insufficiently stiff springs had been fitted to control the tappets, which
were therefore momentarily losing contact with (i.e., 'bouncing' on) the camshaft
at the higher speeds, thus producing an additional impacts, and additional noise.

In both of these examples a change in condition was detected using only
simple measurements of dBA, but it is possible that in both cases changes in level
of one of the octave bands may have been a more sensitive detector.

2.7 MECHANISMS OF NOISE PRODUCTION

Most noise is generated by one of three mechanisms: a vibrating surface, aerodynamic disturbance, or by impacts.

2.7.1 Sound radiation from vibrating surfaces

In many types of machine there is a 'working' force which performs the main
function of the machine, e.g., providing motive power, or some material forming
or shaping action such as cutting, pressing, extruding, etc. Some small fraction of
this force is transmitted to the outer surfaces of the machine, e.g., to the casing of
gearboxes, pumps, fans, motors, etc., causing them to vibrate, and to act, in effect
rather like a loudspeaker, radiating noise to the surrounding air.

In such cases the sound pressure waveform measured by a microphone held very close to the surface may be very similar in pattern to that from a vibration transducer attached to the vibrating surface. The sound waveform will, however, also be influenced by radiation from other noise sources, and by reflections from surfaces in the room, particularly as the distance of the microphone from the vibrating surface is increased. These effects are discussed in more detail in the next chapter.

2.7.2 Aerodynamically generated noise

Such a disturbance occurs when there is movement or flow of air. If air flows smoothly and slowly, along a duct for example, so that the flow is streamlined, or 'laminar' then very little noise is generated. If, however, the smooth lines of flow break up into turbulence, resulting in the creation of vortices, or eddies of air, then the amount of noise increases very markedly.

Such turbulence will occur, even for a smooth uninterrupted flow, if the air velocity increases beyond a certain value, but will occur at much lower flow speeds if there are obstacles in the path of the flow which break up the smooth flow lines. It is the vibration induced in the air by such turbulence that is responsible for the 'swish' of a cane moved rapidly through the air. A similar mechanism accounts for the noise radiated by the blades of a fan, or by the movement of air over a sharp edge, such as occurs in a whistle, or, sometimes, in a machine valve. It is the turbulence generated by the friction between quickly and slowly moving layers of air that causes the noise from the air emerging from the nozzle of a jet, or from a hole in a leaking pressurised pipe or boiler in a factory.

2.7.3 Noise generated by impacts

The noise generated by two impacting objects depends on the momentum of the impact, i.e. on both the relative velocity and the mass of the impacting parts. There is often a secondary source of noise if the impact causes a nearby surface to vibrate, or, ring, as when an object is thrown into a bin or hopper made from mild steel, or onto a sheet steel conveyor chute. Impacts between components in close contact can increase as a result of poor fit, looseness and wear between the two parts, e.g., in bearings and in gears, and this can be detected as an increase in the noise and vibration produced.

2.7.4 Multiple sources of noise in machinery

In many machines there may be several different noise producing mechanisms. In a diesel engine, for example, an important noise producing mechanism is the pressure pulse periodically produced by the combustion process, which generates vibration of the outer surfaces of the machine. However, significant amounts of noise are also produced by the mechanical forces arising from contact between the moving parts, e.g. in bearings, gears and valves. Noise is also produced by ancillary equipment such as generators, pumps and gear-boxes.

Engine noise is often a major part of the total noise from a road vehicle but additional noise producing mechanisms are the impact forces between the tyres and the road surface, and aerodynamically produced noise at the air intake and exhaust ports of the engine. Vehicle body vibration and aerodynamic noise caused by the movement of the vehicle through the air also contribute to the total noise.

In all forms of rotating machinery there will, in addition be out of balance forces which arise because, no matter how well engineered, no machine can ever be perfectly balanced. These forces will produce vibration and noise.

In motors and generators noise and vibration are produced by the electrical and magnetic machine operating forces, in addition to out of balance and mechanical sources of noise. Noise is generated in pumps and fans by aerodynamic and hydrodynamic forces.

In material forming machinery cutting and pressing forces generate noise in saws, drills, grinders, lathes, mills and presses.

In fluid power systems there are two major sources associated with the hydraulic pump - liquid pressure pulsations due to the cyclically varied displacement (e.g., the number of pistons) and the mechanical pump movements; in addition, cavitation noise may occur due to low inlet pressures. Resonating pipes and valve changes also add to the sound generated.

Differences in many of these features due to a fault developing, can, therefore, be detected from analysis of the sound.

2.8 THE EFFECTS OF NOISE ON MAN

Noise can damage hearing, interfere with speech communication, cause disturbance with sleep, cause annoyance and affect the efficiency of the performance of various tasks. There is also evidence that noise is related to the prevalence of certain stress related health effects, and have affects on social behaviour.

2.8.1 Noise induced hearing loss

The damaging effects of prolonged exposure to high levels of noise on human hearing has been known almost from the time of the industrial revolution. Occasional exposure to high levels of noise for several hours can cause a temporary loss of hearing, which recovers after a period of relief from the noise exposure. However, if the level of exposure continues to be repeated, permanent damage occurs to the noise sensitive hair-like cells of the basilar membrane of the cochlea, particularly to the high frequency sensors - see **Figure 2.11**. Noise induced hearing loss is often associated with the condition of tinnitus, a ringing in the ears with no acoustic cause. Extremely high levels of noise, even of short duration, from gunfire or from an explosion, for example, can also cause permanent damage to hearing. (It is thought that the Duke of Wellington permanently lost his hearing due to a gun salute given in his honour!) This is called acoustic trauma.

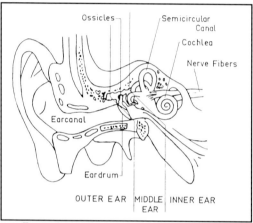

Figure 2.11 - The human ear [Brüel & Kjær]

In order to protect employees from noise induced hearing loss the **Noise at Work Regulations (1989)** require that employees exposure to noise be reduced to below an equivalent continuous noise level of 85 dBA over an 8 hour period each working day. In order to protect against acoustic trauma peak sound pressure of above 200 Pa must be avoided.

It has been estimated that about 15 to 20% of the working population of Europe is exposed to noise levels of between 75 and 85 dBA, and that in the UK somewhere between 0,75 and 2 million employees are exposed to continuous equivalent noise levels of above 85 dBA.

2.8.2 Interference with communication

It is well known that bursts of high levels of noise can sometimes make it impossible to listen to the radio or television, or to continue a conversation or listen to some other audible signal. This is because of the phenomenon of masking, whereby the presence of one sound reduces the audibility, or masks, other sounds. The masking sound causes the threshold of hearing for other sounds to be increased, so that unless the signal level, i.e., the listening level, of the radio or television, for example, is increased, it will not be heard above the masking sound.

For many people the most undesirable effect of noise in their environment, from aircraft or road traffic for example, is the interference with conversation in their homes and their enjoyment of television or radio. At work, in the office noise can interfere with communication and thus lead to mistakes and inefficiencies, and in a factory environment can lead to accidents if warning and alarm signals cannot be heard above the background noise.

The design of public address systems to achieve good levels of intelligibility in conditions where there are high levels of background noise has become of great importance in public places, such as shopping malls, and sports stadia, in order to be sure that emergency alarm messages can be clearly heard and understood.

Masking can sometimes be put to good use, in offices for example, when the introduction of a sound masking system can be used to increase acoustic privacy by masking conversations transmitted between offices, or between offices and adjacent corridors. Masking sounds are also sometimes used by tinnitus sufferers to alleviate the effects of their tinnitus. A masking noise is most effective at masking sounds of the same frequency range, but it is more effective at masking sounds which are above its own frequency range than it is at masking those below this range.

It is the level of noise in the 500, 1000 and 2000 Hz octaves bands which are the most important for speech interference. Noise levels indoors should not exceed 40 to 45 dBA in order not to interfere with conversations, and much lower levels of noise are needed where particularly good listening conditions are required, in classrooms and rooms used for meetings. External noise levels of about 70 dBA and above may cause people to close windows in order to prevent the noise from interfering with listening to speech.

2.8.3 Sleep disturbance

Noise can interfere with sleep in a number of different ways. It can effect the time it takes to fall asleep, and cause awakening. Various levels of sleep have been identified and exposure to noise whilst asleep can lead to changes in sleep patterns, from a deeper to a less deep level of sleep.

Sleep disturbed by noise can lead to changes in mood the following day and to other symptoms such as feelings of annoyance and headaches. These effects can be brought about both by continuous noise and by short bursts of noise. It is the ill and the elderly in society, and shift workers who are most sensitive to these effects. Sleep disturbance starts at noise levels of about 30 dBA for continuous noise, but short bursts of noise above about 45 dBA from individual events, such as passing lorries or aircraft, can also cause sleep disturbance.

2.8.4 Annoyance

This is one of the most widespread and common effects of noise. Annoyance related to noise has been defined as a feeling of resentment, displeasure, discomfort, dissatisfaction or offence that occurs when noise interferes with someone's thoughts, feelings or activities. The amount of annoyance caused depends on a number of non-acoustical factors, as well as on the physical characteristics of the noise. The latter include the level, frequency content and duration of the noise and the temporal pattern of intermittent noise including factors such as the number and frequency of occurrence, and time of day of noisy events. Non acoustical factors related to annoyance include the general feelings and attitudes of the listener towards the source of the noise, what he or she is doing at the time, the degree to which the noise can be controlled, and its perceived avoidability.

The level of the background noise against which the intrusive noise takes place is also an important factor in determining annoyance. The sensitivity of people to noise varies greatly from one individual to another, and even the response of any one individual can vary on different occasions. Using social survey techniques it is possible to measure and relate the average degree of annoyance of a population or community of people to the level of noise, but it is not possible to predict the response of any one individual in that community. An approximate threshold for the onset of annoyance for steady continuous noise is an external L_{Aeq} of 50 dBA, with few people becoming seriously annoyed by exposures less than 55 dBA. The numbers of people seriously annoyed increases for exposures between 55 and 60 dBA, and increases rapidly above 60 dBA. Most people will be seriously annoyed in the daytime and there will be severe disturbance at nighttime by noise exposures of 65 dBA and above.

It has been estimated that in 1994 about 80 million citizens of the European Union about 20 % of the population, live in so called 'black spot' areas where the continuous equivalent noise from road traffic, measured outside dwellings exceeds 65 dBA. This is the level which most people, in social surveys, judge to be unacceptable.

2.8.5 Reduced/improved performance of tasks

Most people would prefer to be in a quiet, rather than in a noisy environment, when trying to concentrate on a task, such as sitting an examination for example, which requires a high level of concentration. However, whilst performing other, more routine tasks, such as household chores for example, a background of speech or music is often preferred. These indications of preference do not, however, indicate that the effectiveness of the task performance will be affected, either one way or the other, by the presence of the noise.

The above examples indicate that the effects of noise on task performance are not straightforward, and many factors are involved. Noise can distract attention from the task, particularly from complex tasks involving mental activity, but, equally, noise can sometimes increase arousal and increase performance. If the tasks rely on hearing speech or other sounds then performance may be adversely affected if the signals are masked by noise. The effect of the noise will depend upon characteristics of the noise such as its regularity, intermittency, information content and its familiarity to the listener.

2.8.6 Stress related effects of noise

Many investigations have shown that blood pressure is higher in noise exposed workers and also in populations living in noisy areas around airports and busy roads than in populations where noise levels are lower. Other studies have shown similar correlations between exposure to high noise levels and certain biochemical changes in blood composition which can lead to ischaemic heart disease and effects on the immune system. The presence of these correlations does not however necessarily prove a direct causal link, and these effects are attributed not to any physical effects caused by the noise but due to the fact that noise can lead to stress, perhaps because of annoyance, or interference with communication, or with sleep.

The fact that leisure activities can sometimes expose individuals to higher levels of noise than even the highest levels of acceptable environmental noise without suffering these effects confirms the view that it is the disturbing and intruding nature of the noise and the stress which it causes, rather than its physical characteristics of the noise which are responsible for most effects on health.

2.8.7 Effects on social behaviour

There is some evidence that noise levels above 80 dBA can reduce 'helping' behaviour, and lead to an increase in aggressive behaviour, and that continuous exposure to high levels of noise may contribute to susceptibility to feelings of helplessness in school children. For most people however the most important effect of noise in the community is interference with rest, recreation and watching television.

CHAPTER THREE

PRACTICAL ISSUES

3. Practical Issues

3.1 INTRODUCTION

In the case of vibration monitoring, the level measured by an accelerometer attached to a machine will not be affected by the proximity and surface finish of walls, floors and ceilings of the room in which it is situated, nor by the operation of other machinery in the same room, provided that the machines are vibration isolated from one another and the surroundings. However, the measured level of noise produced by the machine in the room will be affected by all of these factors.

Many of the practical issues relating to carrying out controlled, accurate and reproducible measurements of noise output from machines are related to the effect of the measurement environment. An understanding of these issues is essential if the condition monitoring engineer is to make reliable and meaningful noise measurements, and, therefore, the reader is encouraged to first assimilate the theory concerning the behaviour of sound waves in rooms discussed in Chapter 2.

This current chapter describes some of the more practical features associated with the measurement of sound waves, and then examines the practical issues related to the use of instruments.

3.2 MEASUREMENT TECHNIQUE

3.2.1 Measurement in enclosed spaces

The effects of reflection and diffraction are discussed in chapter 2 and have a direct influence on the microphone in enclosed spaces. In order to minimise these reflecting and shielding effects on the microphone a hand-held sound level meter should always be held at arm's length away from the operator's body, and pointed towards the machine. Alternatively, the sound level meter may be mounted on a tripod and, in extreme cases, the microphone may be separated from the sound level meter and the observer by a length of microphone cable.

The presence and nature (i.e., whether sound reflecting or absorbing) of any surfaces close to the microphone should always be noted. Several standards on noise measurement specify that the microphone should not be positioned closer than 1 metre from any surface.

Sometimes, when a machine noise contains a pure tone component standing waves can occur in rooms, as a result of interference between waves which are reflected between parallel sets of surfaces (e.g., between opposite sets of walls, or between floors and ceilings). They are characterised by large periodic variations of sound level with position, with maximum levels occurring as a result of constructive interference at positions half a wavelength apart, with positions in between, where the noise level is a minimum, resulting from destructive interference. The significance for noise monitoring is that the sound pressure level from the machine may be significantly different from that which would have been measured at the same position in a room where the standing waves were absent.

a. The effect of background noise

As mentioned in Chapter 2, Section 2.3.4, a microphone placed close to a machine will respond to background noise at a measurement position as well as to the noise from the machine. It is, therefore, an important general rule of good practice in noise measurement that the background noise should always be at least 10 dB below the noise which is being measured, in order that the background noise shall have a negligible influence on the measurement result.

It is therefore good measurement practice when measuring the noise from a source such as a machine to repeat the measurement with the machine switched off, in order to measure the background noise as well. If the background noise level is not at least 10 dB below the level of the machine noise then it is possible to attempt to correct the measured level to take into account the effect of the background noise, using the chart given in **Table 3.1**, or the formula:

$$\text{Corrected level} = 10 \log (10^{L_M/10} - 10^{L_B/10}) \qquad\qquad (3.1)$$

where L_M = the level of the machine noise measured in the presence of the background noise
L_B = the level of the background noise, with the machine switched off.

Measured noise levels less background noise level - dBA	Amount to be subtracted from measured level - dBA
3	3
4 to 5	2
6 to 9	1
10 or more	0

Table 3.1 - Correction chart for background noise

Note that if measurements are being made in octave or third octave bands then the background noise must also be measured in the same way.

The above procedures can only provide an approximate correction for the effects of background noise, and therefore the preferred procedure would always be to arrange for the machine noise to be measured again, when the background noise is lower.

b. Effect of room characteristics

The effect of reflection from a room's surfaces in producing a reverberant sound field in addition to the direct sound field was discussed in chapter 2.

3.2.2 The effect of distance on the measurement of sound

If the sound pressure level is measured at some distance away from the source, in a far field position, say 10 m for example, it will be possible to predict the sound pressure level at some other distance, say 30 m for example, using the '6 dB per doubling of distance' rule. Such predictions can never be made from near field measurements. **Figure 2.9** shows a real noise source and the contours which represent equal sound pressure levels, the dB decreases 6 dB for each doubling of the distance from the source.

Unfortunately, it is not possible to precisely define the boundary between the near and far field regions, except for idealised sources of simple geometry. The extent of the near field region does, however, depend upon both the maximum dimensions of the source and on the wavelength of the sound, and ideally, a receiver should be at a distance from the machine which is both several wavelengths and several machine dimensions away.

The significance of the near/far field concept for condition monitoring is that the sound pressure level changes much more rapidly with position in the near field than in the far field. Therefore, much greater care must be taken when using a microphone close to a machine to ensure that its position is accurately measured, and can be repeated precisely, than when the microphone is situated in the far field of the machine, when a difference of a few millimetres will not be so critical.

The important consequence for condition monitoring is the measured sound pressure level from a machine will depend on the characteristics of the room, as well as on the sound power level of the machine, unless the noise from a machine is measured close enough to the machine for the direct sound to predominate.

3.3 STANDARD ACOUSTIC TEST ENVIRONMENTS - ANECHOIC AND REVERBERANT ROOMS

In most spaces both direct and reverberant sound will be present, in different proportions, depending on the position within the room, relative to the source. These are called semi-reverberant rooms.

Anechoic and reverberant rooms are standard acoustic test environments in which one or other of these two sound fields is entirely dominant, and the other is negligible by comparison. They can be used to measure the sound power level of machines.

An anechoic, or dead, room is one in which all the room surfaces are completely sound absorbing, so that there is no reverberant sound, and only direct sound is present. This is achieved by covering all the room surfaces with wedges of sound absorbing material. Sometimes the floor is untreated and is sound reflecting, in which case the room is called a semi-anechoic room.

A reverberant, or live, room is one in which all the room surfaces are hard and sound reflecting and where sound energy in the room is distributed evenly throughout the space, so that there is only reverberant sound. This is achieved by making all the room structural surfaces hard and sound reflecting, and using sheets of material within the space to act as diffusers of the sound and break up any standing wave patterns which might otherwise be formed. Sometimes the walls are slightly skewed, i.e., made to be non-parallel. The type of pattern for the sound pressure in various types of rooms is shown in **Figure 2.10**.

3.4 THE SPECIFICATION OF NOISE EMISSION FROM MACHINERY

The amount of noise produced by a machine, called its 'noise emission', may be specified in one of two ways:

- either, as a sound pressure level measured at a known distance from the machine,
- or, as a sound power level.

The use of sound pressure level at a known distance has the advantage of simplicity, being easily understood, and being easy to measure, requiring only the use of a sound level meter. The big disadvantage with this method, however, is that the sound pressure level will depend upon the environment surrounding the machine, as discussed earlier. Therefore unless the environment can be standardised the result may not be applicable to situations when the machine is located in another room, with different acoustic characteristics.

The use of sound power level overcomes this difficulty, because it is a value that relates only to the machine, and is independent of the acoustic environment in which the room is situated. A disadvantage of using sound power level is that it is not so readily understood, since sound pressure levels, which are more directly related to what people hear, have to be calculated from the sound power level. Another disadvantage is that sound power level is more difficult to measure than sound pressure level.

There are three standard methods for determining the sound power level of a machine. The first two require that the machine be placed in a standard test environment, either an anechoic or a reverberant room. From measurements of sound pressure level, using a sound level meter, the sound power level may be estimated. The third method requires the use of a sound intensity meter. This method has the advantage that it allows the sound power level of the machine to be measured in situ, without the need for the machine to be moved to a specialist test environment. The use of the sound intensity meter is outlined in Section 2.4.6.

There are several series of EC Regulations on machinery noise which require that the noise emission of many different types of machinery be published by manufacturers and/or suppliers of the machine. The various test methods are specified in the Regulations - these often require the measurement of sound power level.

It is, of course, essential that in any noise emission test, whichever measurement method is used, the machine operating conditions are precisely specified to such a degree that would allow the test to be reliably repeated.

3.5 THE USE OF EQUIPMENT

With the development of digital signal processing in recent years sound measurement equipment has become ever more complex and automated. No equipment, however, is foolproof, and without an understanding of simple acoustic measurement principles it is quite possible to achieve meaningless results even with the most sophisticated equipment.

It is important when using sound measuring equipment to listen and look, and with a little practice it is usually possible to correlate changes in measured sound levels with what is heard. This gives confidence that the equipment is working properly, and, most importantly, gives helpful clues that may warn that all is not well and that the meter has developed a fault, or is being influenced by factors other than noise.

Some modern sound measurement equipment can be complicated to use, and the manufacturer's handbook should always be read carefully before use and stored in a safe place for future reference if necessary.

3.5.1 Pre-measurement checks

The next step is to switch the equipment on, and to carry out simple checks to ensure for example that the meter responds to variations in level of simple sounds such as voice, hand-claps, etc. Some battery-operated equipment has indicators of battery condition, and these should always be checked both at the beginning and at the end of every measurement session, and spare batteries carried with the equipment.

3.5.2 Microphone care

All types of microphone used for sound measurement are extremely delicate and must always be handled with extreme care. In some equipment the microphone is always stored separately from the rest of the instrument and the first step in setting up the equipment, before switching on, is to carefully attach the microphone to the meter. A protective grid is usually positioned over the microphone diaphragm, and this should only be removed, with great care, when absolutely necessary, for example, to inspect the diaphragm for suspected damage.

Some equipment can be adapted to use either condenser or electret microphones. The major difference is that the condenser microphone needs to be supplied with a polarising voltage, commonly 200 volts, whereas the electret, also called the pre-polarised microphone, does not. The appropriate polarising voltage, including the zero volts (for a pre-polarised microphone) must be selected from the instrument control options.

3.5.3 Calibration

a. Microphones

Microphones are usually calibrated by the manufacturer for use in different types of sound field. There are commonly three configurations:

- free-field,
- pressure response and
- random incidence.

Free-field microphones are the most suitable for situations where the noise is mainly due to direct sound from a source. The calibration of free-field microphones incorporates a built-in correction factor to allow for diffraction effects by the microphone, so that the measured sound pressure level corresponds to that which would occur in the absence of the microphone. This type of microphone is usually calibrated for zero degrees angle of incidence, i.e., for sound incident normally on the diaphragm, and the microphone should therefore be pointed directly at the sound source.

Pressure response microphones are calibrated so that they measure the sound pressure acting on the diaphragm, including diffraction effects. This type of microphone is usually calibrated for 90° angle of incidence, i.e., for sound 'grazing' over the microphone. Pressure response microphones should be used when it is necessary to remove the sound pressure level acting on a surface, e.g., a wall, or duct, or pipe, in which case the microphone is mounted flush with the surface.

Random incidence microphones should be used indoors where the sound field is mainly reverberant. The difference between the three types of calibration becomes important at high frequencies, above 2 kHz. Some sound level meters have electronic correction to allow for any type to be used, in which case the meter must be switched to the appropriate selection. **Figure 3.1** illustrates these three configurations.

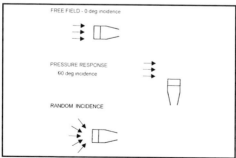

Figure 3.1 - Three calibration configurations of microphones

b. Complete system

The next step is to calibrate the equipment. All good quality equipment is designed to be used in conjunction with an acoustic calibrator, which is a device which, when fitted closely over the microphone of the sound level meter produces a precisely known sound level. If there is a slight deviation between the calibration level and that indicated by the sound level meter this can be corrected to ensure that the meter always reads accurately.

The meter should be calibrated at the beginning and end of every measurement session. If any extension cables are fitted between the microphone and the meter, in order to measure in confined or inaccessible places for example, then the calibration should be carried out with these in place, i.e., it is always the complete measurement system which is calibrated.

An alternative form of calibrator is also available, which consists of an electrostatic grid (called an electrostatic actuator) which fits over the diaphragm. A voltage signal applied to the grid provides an electrostatic force to the diaphragm, which simulates the effect of a sound pressure. This type of calibration can be performed automatically, for microphones which are remote or inaccessible.

The sound level meter and the calibrator should be re-calibrated by an accredited calibration laboratory at regular intervals of at least once every two years.

Figure 3.2 - 475/485/801M dynamic response calibrator system (Beran Instruments)

Figure 3.2 shows an automatic microphone calibration system. Calibration is performed utilising an acoustic coupler with a reference microphone calibrated to national standards; with this configuration traceable calibration is achievable. The calibration system automatically measures the output of the microphone and pre-amplifier, and then applies an insert voltage in accordance with IEC1094-1 (1992) and ANSI S1.10-1966 (R1976). For calibration up to 100 kHz the electrostatic actuator technique is utilised. All calibration results are stored into an internal database for certificate production and further analysis or trending.

3.5.4 Use of windshield

It is good measurement practice to fit a foam plastic windshield over the microphone to reduce the effect of any air moving across the face of the microphone, provided that the air speed is not greater than a few metres per second (the manufacturer's handbook should be consulted for exact details). The use of a windshield also provides some physical protection against accidental knocks.

3.5.5 Selection of measurement parameters

The appropriate measurement parameters should be selected, e.g., dBA or octave bands, Fast, Slow, Peak or L_{eq}, and the appropriate instrument range setting selected to ensure that the measurement signal lies comfortably within the measurement range, so that the signal is neither overloaded nor lost within instrument noise. Many modern instruments have automatic range selection and/or overload and underload detection indicators to assist in this process.

3.6 SELECTION OF MEASUREMENT POSITION

The selection of the measurement position has to be considered with some care, taking into account the issues raised earlier in this chapter, i.e., the effects of background noise, nearby reflecting surfaces, direct and reverberant and near and far sound fields, the presence of standing waves, etc.

The overriding practical consideration is that the measurement conditions shall be repeatable, and therefore whatever microphone location is selected its position must be measured accurately and recorded so that it can be replaced in exactly the same position relative to the machine, for future comparative measurements. Note that it is necessary not only to register the position of the microphone but also its height above the ground, and its orientation, i.e., the direction in which it is pointing. Some microphones are calibrated for free field use and at normal incidence, in which case they should be pointed with their axis directly at the machine. Others are calibrated for 90 degree incidence or normal incidence.

Very often the need for the measured machine noise to be at least 10 dB above background level will determine that the microphone be located very close to the machine. **Figure 3.3** shows an example of a site plan.

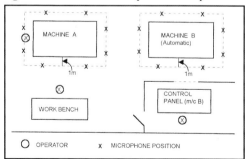

Figure 3.3 - Example of a site plan for noise measurement

The need for the measurement conditions to be repeatable requires that the machine operating conditions, i.e., speed, load, etc., must also be noted, and must be repeatable.

3.6.1 Environmental influences

The ideal sound measuring instrument should be sensitive to sound alone, and completely insensitive to, and unaffected by, variations in other aspects of the environment. Although modern equipment is usually very satisfactory in this respect the meter can be affected by environmental influence such as moisture and humidity, wind, vibration, and by electric and magnetic fields. The equipment manufacturer's handbook should give guidance on these matters, but careful observation by looking and listening and correlating what is seen on the meter with what is heard is invaluable in detecting any non-acoustic influences on the meter reading.

a. Wind

In most cases the use of a windshield will provide adequate protection against gentle air movements but not against the influence of air flows of several metres per second.

b. Moisture

Condenser microphones are particularly sensitive to the effects of moisture and humidity. The danger arises because the air gap between the microphone diaphragm and its backing plate, across which there is a polarising voltage of up to 200 volts, is only a few microns. Dry air is a very good electrical insulator and well able to sustain this high level of electric field, but water is a relatively good conductor of electricity. Therefore if droplets of water condense in this gap, arcing or sparking occurs, which may result in expensive damage to the microphone diaphragm, and incorrect measurement results. Therefore it is advisable not to take sound measurements outdoors in the rain, or indoors where the microphone can be affected by sprays of water.

Non-polarised microphones do not suffer from this susceptibility to moisture, but even when using equipment fitted with this type of microphone it is common practice not to take measurements in wet conditions, unless there are special reasons for doing so, because noise from water splash affects the measurement results. There is also the danger of water seeping into the 'electronics' of the instrument and causing damage.

c. Electromagnetic radiation

It is possible that the sound level meter could be affected by electromagnetic radiation when used in close proximity to transformers, motors, etc., in which case the extraneous signal is likely to occur at 50 and/or 100 Hz frequencies.

d. General

Sometimes it is possible to detect any non-acoustic influences by repeating the sound measurement at a position where the noise level sounds similar to the ear, but where the suspected influence (wind or electric field) would be reduced or absent.

3.7 THE ACCURACY OF NOISE MEASUREMENTS

The accuracy of sound level measurements depends upon:

- the accuracy of the sound level meter,
- the way in which it is used and
- the variability of the noise being measured, which usually contains some random fluctuation.

The first of these factors will be discussed in Chapter 4, but it is usually the second two factors which have the greatest influence on the overall accuracy of the result. An indication of this accuracy can be obtained by repeating the measurement and looking at the variability of the result. Unless the measurement is taken in closely controlled laboratory conditions an accuracy of ± 1 dBA is the very best that can be achieved, with variations of ± 2 or 3 dBA quite likely. Larger variations can occur in octave or third octave band measurements, particularly in the lower frequency bands.

Because of this, noise measurements should always be rounded and quoted to the nearest decibel, even if digital measurement equipment, or prediction calculations indicate a result to the first or second decimal point.

3.7.1 Post-measurement checks

At the end of every measurement session a series of checks should be carried out to ensure that the equipment is in working order and that all correct procedures have been followed. This could include: re-calibration, battery checks, checks on selection of correct meter settings, etc., and that all relevant details have been recorded before leaving site. A measurement checklist is useful.

3.8 MEASUREMENT REPORT FORMS

All relevant details should be recorded on an appropriately designed measurement report form including the items shown in **Table 3.2**.

Subject	Items	Comments
Basic information	Date, time, location, operator's name, etc.	
Noise source	machine type, operating conditions.	e.g., speed, load, feed rate, feed material, etc.
Noise measurement	Measurement parameter, measured values, type of equipment, measurement position.	Including microphone and calibrator.
	Details of measurement checks	Calibration, battery, windshield, etc.
Measurement environment	Room size and position of machine within the room, proximity of sound reflecting surfaces close to microphone, background noise level, nature of room surfaces and any other surfaces close to the microphone.	Usually a sketch plan of the site is useful (see example in **Figure 3.2**).

Table 3.2 - Reportable features

3.9 SPECIALIST NOISE MEASUREMENT TECHNIQUES

The measurement of sound 'intensity', as against sound 'pressure', (see Section 2.4.6) provides the ability to identify the direction of flow of acoustic energy, as well as measuring its magnitude. This is the basis of many of the applications of the sound intensity meter.

One of these applications is the measurement of sound power emitted by noise sources. Using the acoustic intensity meter it is possible to measure the sound power level of a noisy machine in situ, in the presence of background noise from nearby machines and also without the need for a specialist acoustic environment, i.e., an anechoic or a reverberant room. Other applications include the location and identification of noise sources and the detection of leaks and weak link areas in the transmission of sound through panels and partitions.

It has been shown above that two of the major problems encountered when measuring machine noise are the separation of the noise radiated directly from the machine from that reflected from the room surfaces, and from background noise from other machines. The use of sound intensity measurement is one technique which can sometimes be used to overcome these difficulties.

If the noise is transient, or consists of a series of transient events, such as impulsive noise from a press, for example, then it may be possible to use specialist transient analysis techniques to separate the direct and reflected waveforms in the time domain. In effect, these techniques 'gate out' the reflections, leaving only the direct waveform, which is analysed and measured.

For continuous, non-transient signals there are a variety of signal processing techniques available which are designed to enhance signals from a background of noise (meaning non-signal in this context). These include so-called correlation techniques.

3.10 SYSTEMATIC APPROACH

The work described by **Worley (1990)** related to noise control. In other words, it was developed in order to reduce the noise levels in an industrial environment for the benefit of the operators. However, the same systematic approach is very valuable for the engineer wanting to determine the source of an unusual sound.

The general idea of the approach is a series of questions, where if the answer is 'yes' the practitioner is directed to a more specific diagnostic test. The following is adapted from **Worley (1990)** but is specific to condition monitoring of the machinery or system:

A. Noise (rms) versus load

        Loudest when on load   ☐

        Loudest when not on load   ☐

        Other .... ☐

B. Noise appears due to <u>impacts</u>

        (if 'No' move to C)

        On machine parts -

guards	☐	1, 2
working forces	☐	1, 3
clutch	☐	1, 2, 3
indexing mechanism	☐	1, 3
punch breakthrough	☐	1,3
other	☐	

Component or scrap handling -
       on chute/track /conveyor   ☐   1, 3, 4
       on hopper   ☐   2, 3
       on bin or machine   ☐   1, 2, 3
       component on component   ☐   1, 3
       scrap on scrap   ☐   1, 3
       bar on stock tube   ☐   2, 5
       other ....   ☐
In hydraulic system -
       pipes, valves   ☐   7, 9
       load related   ☐   2, 6
       (loudest part ....)
Other   ....   ☐

C. Noise contains tones
    (if 'No' move to D)
    Likely to be from mechanical transmission
       belt drive   ☐   9, 11
       gearbox   ☐   8, 9, 11
       bearings   ☐   10
    Likely to be from electrical power system
       frequency converter   ☐   2, 8
       other ....   ☐
    Likely to be from hydraulic system
       pumps, pipes, valves   ☐   7, 9
       other ....   ☐
       load related   ☐   2, 6
       (loudest part ....)
    Likely to be from cutting action
       tool chatter   ☐   2, 7, 12
       high speed cutter   ☐   2, 10, 11
       workpiece vibration   ☐   3, 8
       obstruction near cutter   ☐   11
       tool vibration   ☐   12, 13
       vibration in machine frame   ☐   7, 14
       other ....   ☐
    Likely due to air flow
       fan noise   ☐   2, 7, 15
       flow over sharp edges   ☐   16
       other ....   ☐
    Other causes ....   ☐

D. Noise considered due to air flow
    (if 'No' move to end)
    Likely to be due to air release (intermittent)
                component ejection   ☐   1, 2, 6
                pneumatic exhaust   ☐   1, 2, 6
                swarf removal   ☐   1, 2
                component drying   ☐   1, 2
                other .... ☐
    Likely to be due to air release (continuous)
                component ejection   ☐   2, 9
                component counting   ☐   2, 9
                leaks   ☐   9
                other .... ☐
    Likely to be due to ventilation or air extraction
                fan noise   ☐   2, 7, 15
                constrictive flow   ☐   16
                other .... ☐
    Any other source
                e.g., compressor   ☐   17
                other .... ☐

The diagnostic tests are given in detail in **Worley (1990)**, but briefly the
following are the general ideas of the 17 tests, slightly updated:

Test 1 - carefully listen, preferably at low speed.
Test 2 - isolate suspected parts of machine to run them or avoid them.
Test 3 - strike all suspected parts with a hammer, one after the other.
Test 4 - vibrationally damp suspected parts by light finger pressure.
Test 5 - note the different aspects of the machine cycle.
Test 6 - endeavour to run the automatic cycle manually.
Test 7 - damp out or remove noise radiators. Use hand-held analyser.
Test 8 - determine rotational speed which affects the response.
Test 9 - remove covers to avoid noise reflection (note safety requirements).
Test 10 - try a hand-held analyser on bearing housing.
Test 11 - determine which rotational components (speeds) modify noise.
Test 12 - examine for poor surface finish on machined components.
Test 13 - with great care try to damp saw blade with long wooden rod.
Test 14 - damping or loading of each panel should identify the fault.
Test 15 - switch off fan and note noise level as the rotation slows.
Test 16 - remove each sharp edge, one at a time.
Test 17 - compare idle with full speed noise load. Look at noisiest loads.

CHAPTER FOUR

EQUIPMENT
AND
INSTRUMENTATION

4. Equipment and Instrumentation

4.1 INTRODUCTION

The aim of this chapter is to provide information to the condition monitoring engineer about what equipment is readily available for the measurement of machine, system, environmental or occupational noise and for building acoustics measurement.

The introduction of large scale integrated circuits, microprocessor technology and digital signal processing over the last few years has meant a very rapid rate of change in the design of sound level meters and other noise and vibration measuring equipment.

There is a bewildering range of hand-held sound level meters on the market costing from a few hundred pounds to several thousands pounds (currently around £200 to £5000). They may be classified by the range of functions they possess, i.e., what they can measure and also by their type according to British and International standards which set down, among other things their limits of accuracy.

The different types in current use range from completely analogue instruments to the 'virtual instrument' consisting of a microphone plus minimal signal conditioning linked to a notebook PC. In between there are other instruments varying from:

- digital display and some digital processing but in which much of the initial signal conditioning and processing is performed using analogue methods, to
- almost entirely digital devices with their own built in microprocessors, dedicated to specific types of analysis (variable in some cases through 'plug-in' ROM modules), or capable of being linked to computers for further 'post-processing'.

4.2 THE BASIC SOUND LEVEL METER

A sound level meter consists of a microphone, various stages of signal conditioning and processing followed by an output or display of the required measurement
parameter. A simplified block diagram of the contents of a typical basic sound
level meter, which is suitable for measuring fairly steady, non-impulsive noises,
is shown in **Figure 4.1**. Each feature is explained later.

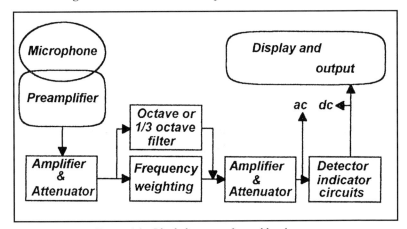

Figure 4.1 - Block diagram of sound level meter

4.2.1 The microphone

The microphone converts a fluctuating sound pressure signal into an electrical
signal in which the variations of voltage mimic identically, or are analogous to
the variations in the sound pressure.

   If, for example, the sound is a 1 kHz pure tone, as used for calibration, then
the microphone produces a voltage which varies sinusoidally with a frequency of
1 kHz. The voltage amplitude should be proportional to the sound pressure amplitude.

   The microphone design is highly important and, possibly, the most expensive part of the whole instrument since any deviation in the conversion process at
this stage cannot be corrected later by electronics. The microphone circuit is required, ideally, to match the performance of the human ear in the range of sound
pressures and frequencies to which it can respond.

   Microphones are discussed in more detail in section **4.7** but one example,
built into a sound level meter is shown in **Figure 4.2**.

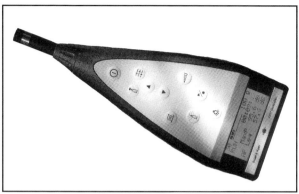

Figure 4.2 - The type 2237 sound level meter
[Brüel & Kjær]

4.2.2 The preamplifier and amplifier stages

The voltage signal from the microphone then undergoes several stages of conditioning and processing before it is in suitable form for the output of a display device. These stages include 'pre-amplification' and amplification.

The purpose of the preamplifier is not so much to amplify the signal but to match the high output impedance of the microphone to the lower impedance of the next stage for amplification to the required magnitude. Because the presence of a cable connecting microphone to preamplifier would load the microphone to an unacceptable degree, the preamplifier is usually built into the microphone cartridge. Thus any microphone extension cables, used to separate the microphone from the sound level meter, or other measuring instrument, must carry the power from the instrument to the preamplifier, as well as the signal from the microphone to the instrument.

The amount of amplification of the signal has to be adjusted according to the level of the signal emitted by the microphone. This is undertaken using a bank of signal attenuators which are controlled using the level range control switch of the instrument.

4.2.3. Frequency weighting networks

The microphone signal, amplified to a suitable level, is led to the weighting networks. Whilst all sound level meters will be fitted with the A-weighting, those used in machine monitoring will also have the option of the C-weightings.

Perhaps, surprisingly, not all sound level meters can measure the unweighted level, but where this is so the measurement is often called the LINEAR level, meaning, on this occasion, "unweighted", and is designated dB(LIN). The unweighted value is used only occasionally for measurement purposes but it is a very useful facility in a sound level meter because it allows an excellent, unweighted, sample of a noise to be captured using a tape recorder, for a variety of subsequent analyses, or for replay through a loudspeaker.

4.2.4 Frequency analysis filters

As an alternative to the weighting networks, some instruments may have the facility for banks of octave or third octave filters to be attached to the instrument, allowing a frequency analysis of the noise to be performed in real time.

In most sound level meters the octave filter sets are arranged in parallel with the weighting networks, so that these various possibilities are alternative options, i.e., one can have an A-weighted level or an octave band analysis, but not both at the same time. In some sound level meters, however, they are in series, i.e., with the weighting networks followed by octave band filters, thus allowing the possibility of an A-weighted octave band analysis.

4.2.5 The detector and indicator circuits

These have four main functions:

- to produce a signal which is proportional to the root mean square (rms) sound pressure;
- to average the variation in the rms signal over an appropriate averaging time (Fast or Slow);
- to deliver the appropriately weighted and averaged signal to the output or display device, and
- to produce a logarithmic value of the signal so that the output/display meter will be read in decibels.

The Fast and Slow averaging times were devised for use with analogue displays, i.e., a needle moving across a voltmeter scale. Usually, when switched to FAST the meter needle will fluctuate over a range of 1 or 2 decibels, even when measuring an apparently steady noise such as from a fan, or vacuum cleaner, or electric drill. These fluctuations may be evened out, visually, and an average value taken. The SLOW mode is intended for use when fluctuations of the meter needle in the FAST mode are more than about 4 dB, making it difficult to judge an average value. Switching to SLOW reduces and slows down the fluctuations and makes it much easier to read the meter.

When measuring steady noises the sound level reading, averaged over a few seconds, should be the same whichever setting, FAST or SLOW, is used. The FAST mode is more suitable when it is required to follow fairly rapid changes in the noise level, i.e., it is required to measure the maximum noise level produced by a car or lorry as it passes by.

4.2.6 The output/display

In some sound level meters an analogue display device is used, such as a moving needle or pointer on a scale. In recent years, however, a digital display is more common, sometimes together with a pseudo-analogue display such as a moving column indicating how the measured parameter is varying.

Where a scale is used, it is graduated in decibels having been calibrated to read directly in decibels relative to 20 µPa. Only a limited decibel range (e.g., maybe, 20 dB) is displayed on the meter scale, and, in order to cover the entire measurement range of the instrument, it is necessary to operate a range selector switch, which controls the gain of the amplifier circuits. **Figure 4.3** shows a variety of displays used on hand-held units.

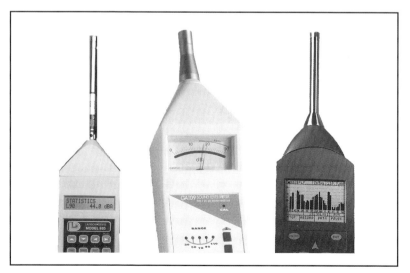

Figure 4.3 - A variety of displays [Larson Davis, Castle and CEL]

4.2.7 Digital sound level meters

All of the above stages may be carried out by analogue signal processing methods. An analogue to digital converter (ADC) in the sound level meter samples, digitises and stores the signal for subsequent digital processing.

The use of digital signal processing enables the technology of the computer

Figure 4.4 - The CR 704B general
purpose type 2 sound level meter
[Cirrus Research]

to be utilised inside the sound level meter, and greatly facilitates processing where statistical calculation is involved, as in the determination of such parameters as, L_{aeq}, L_n, L_{max} and L_{min}.

In the latest digital sound level meters (**Figure 4.4**) the only analogue processes which are retained are the preamplifier and amplifier stages. Both frequency weighting and rms averaging are performed digitally. The digitised waveform signal may also be frequency analysed using digital filters, so that the variation in frequency spectrum with time may be monitored in real time and time variation measures such as L_{10}, L_{eq}, etc., are also available in octave or third octave bands.

4.2.8 Additional sound level meter features

Display Developments in electronics have produced sound level meters with a wide variety of displays (see **Figure 4.3**). Improvements in rms circuits have made available meters with a greatly extended indicator range of up to 140 dB on the meter scale as compared with 20 dB on the older types. This makes the meter much easier to use since there may only be the need for three, or perhaps just two, switchings to cover the whole range. Some meters even have automatic range changing. *Maximum level indicators* Analogue display meters may also be fitted with a HOLD facility which 'freezes' the meter needle at its highest position during a noise measurement, making it much easier to measure maximum readings. The meter needle is released by a RESET button after each measurement.

Digital display options Digital displays are updated after a certain refresh time, usually one second. In addition to the instantaneous sound pressure level at the moment of refresh a number of other measures are often available: for example, maximum and minimum values within the last second, and the maximum and minimum values since the start of measurement.

Since a digital instrument will usually have integrating, statistical and peak measuring facilities, a large number of measurement parameters may be available from a single measurement and the user must, with the use of the instrument handbook, be sure of exactly what is being measured.

Outputs The sound level meter may be fitted with output sockets to allow the signal to be taken to other devices for various types of display, analysis or recording. The output may be taken from different stages of the signal processing chain, thus:

1. An *ac output* taken after the weighting network but before the detector and indicator circuits. The signal at this stage, although weighted still contains the sound pressure waveform information, and may be fed into a tape recorder, headphones or loudspeakers, or to an oscilloscope for visual examination of the waveform.
2. A *dc output* taken after the signal has been through the rms circuits. This signal is suitable for feeding into a level recorder or other form of chart recorder to enable a permanent paper record to be made of the noise level and its variations.
3. In a digital meter it may be possible to take the *digitised signal* to a printer for display, or to a computer for storage and further analysis. **Figure 4.5** shows a typical printed output.

Overload/Underload indicators These are flashing light indicators fitted to some sound level meters particularly those with IMPULSE measuring facilities where there is no automatic adjustment. They flash when the amplifier circuits are being overloaded by the signal and the attenuator settings have to be changed to ensure an accurate measurement. Overload indicators are particularly useful in detecting sharp transients for which a meter needle is unable to respond.

Some instruments are also fitted with 'Underload Indicators'. Again the flashing light is a warning to the user that that the display range needs to be adjusted.

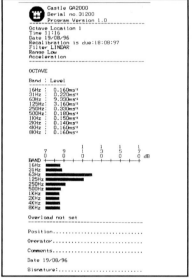

Figure 4.5 - Typical printer output [Castle]

Alternative inputs Some sound level meters allow signals from external devices to be fed into the input amplifiers of the meter instead of via the microphone. Examples are signals from accelerometers and other types of vibration transducer or from a tape recording, allowing perhaps a signal recorded from a fairly basic type of sound level meter to be processed and measured by equipment with more sophisticated analysis facilities.

4.2.9 Accessories

Microphone extensions The microphone may be separated from the body of the sound level meter by a flexible microphone extension bar, or 'goose-neck'. This may be helpful when it is required to reduce the effects of reflection from the body of the meter, the observer or any other nearby surface, or when measuring in inaccessible positions. The microphone may be located even more remotely from the meter by using a microphone extension cable and by mounting the microphone and meter separately on tripods.

Windshields Windshields are available which fit over the microphone and produce only minimal changes to the sound level reading, but which protect the microphone from distortion caused by the passage of wind over the microphone (**Figure 4.6**). They are effective up to certain wind speeds, quoted by the manufacturer. It is good practice to consistently use windshields since they are always required for outdoor measurements in case of sudden gusts of wind, even on a calm day, and are often required indoors as well where there may be considerable air movement caused by natural draughts, forced ventilation and by cooling fans present in many items of machinery. The penalty for not using a windshield may be an invalid measurement. In addition, the windshield may provide some protection to the microphone against accidental knocks.

Figure 4.6 - A protective microphone windshield [Castle]

4.3 IMPULSE SOUND LEVEL METERS

When it is required to measure very sharply varying impulsive sounds, such as thumps, bangs and clatters produced by sudden impacts, or noise from explosions or gunfire, then neither the FAST nor SLOW mode is suitable or adequate. For measurement of these noises a sound level meter with detector and indicator circuits capable of dealing with impulsive sounds and producing a signal proportional to the peak value of the waveform signal, rather than to its rms value is required. An example of this type of meter is shown in **Figure 4.7**.

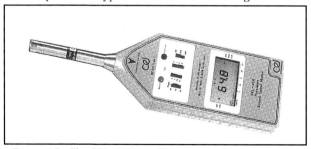

Figure 4.7 - The CEL-414 impulse sound level meter [Casella]

4.4 INTEGRATING SOUND LEVEL METERS

Sound level meters which are able to measure the equivalent continuous sound level of a time varying noise, L_{Aeq}, are called integrating meters **Figure 4.8**. The name arises from the mathematical definition of L_{Aeq}, which involves the integration of the varying sound pressure squared with respect to time, over the measurement period. Integrating meters are used for measuring fluctuating, intermittent and impulsive noises, as well as being convenient for relatively steady noises where the sound level fluctuations are too great to allow accurate sound level readings to be taken even using the SLOW time weighting.

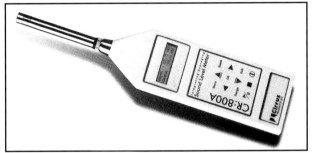

Figure 4.8 - The CR 800A integrating sound level meter [Cirrus Research]

Some meters have the facility to measure sound exposure levels (SELs) which is particularly useful for measurement of noise from single events such as aircraft flyover, trains passing by, or bursts of process noise. The SEL is the steady level which, over a one second period would contain the same amount of energy as the noise event under consideration.

L_{eq} is almost always measured as an A-weighted value but it is possible to use other frequency weightings or to have linear octave-band L_{eq} values. The same is true of SEL.

Measurement of L_{eq} may be required over periods ranging from a few seconds to several hours. The measurement period may be under the control of the operator, using a RESET button, or through the use of pre-set measurement intervals such as 1 minute, 5 minutes, 15 minutes, 1 hour, etc. Some instruments may be programmed to start measurements at pre-determined times, or when noise levels reach certain pre-determined trigger level. The meter may be fitted with an indicator of elapsed time since the start of integration, and with a PAUSE button, to enable a burst of unwanted noise to be excluded from the measurement, for example to exclude noise from an aircraft during measurement of traffic noise.

Because the noise level may vary over a wide range during the measurement of L_{eq} integrating sound level meters are required to have wide measurement ranges. It is therefore particularly important that they are fitted with overload indicators to warn if the sound pressure has exceeded the measurement range during the integration period.

4.5 PERFORMANCE SPECIFICATION FOR SOUND LEVEL METERS

British Standard BS EN 60651:1994 describes the specification of sound level meter performance. This is based on four characteristics of sound level meters:

1. directional characteristics
2. frequency weighting characteristics
3. the characteristics of detector and indicator circuits
4. sensitivity to various environments.

Accuracy Four types of sound level meter are designated: types 0, 1, 2 and 3.

- *Type 0* is the most accurate and is intended for laboratory use only.
- *Type 1* is slightly less accurate, and is intended for both laboratory work and for the most accurate type of field work.
- *Types 2 and 3* are intended for field measurements only.

The specification in the standard is, the same for each type of meter, but the permitted deviations from the specification vary for each type, being greatest for the least accurate type 3 meters and least for the most accurate, type 0 meters.

Directionality For most situations, a sound level meter, ideally, should be omnidirectional, i.e., should respond equally to sound arriving at any angle of incidence. However, this is not so in practice, because of the effects of diffraction. The deviation from this ideal 'omnidirectionality' increases with frequency, and also depends on the type number, being smallest for type 0 and greatest for type 3 meters.

Frequency weightings The frequency characteristics of the various weighting networks (A, B, C and D) are given in BS EN 60651. The allowed tolerance on these values depends on type number, being smallest for type 0 and greatest for type 3 meters.

Linear or unweighted SPL The standard does not cover LINEAR or unweighted measurements. This means that different sound level meters may operate over different frequency ranges, and tolerate different departure from an ideal flat frequency response. For this reason the use of C weighting, which is the standard frequency weighting which is closest to linear, is increasingly being specified in standard measurement procedures, particularly for peak values, and being included in sound level meters, in addition to the A-weighting.

Time weightings The time averaging in the sound level meter is performed by exponential averaging circuits based on RC (resistor-capacitor) networks. A continuously varying 'running' average value is displayed, with the contributions of the past instantaneous values of the signal to the currently displayed average value reducing exponentially according to the elapsed time The time constant for the exponential decay process is the time taken for the value of the decaying parameter to become reduced to 1/e of its initial value (where 'e' is the exponential number, 2,718, and 1/e represents a fall to 37% of the initial value). This time constant is 1/8 second for the FAST circuit and it is one second for the SLOW network.

The performance of these circuits is specified in the Standard in terms of how quickly the meter needle rises when a sudden burst of pure tone signal is received.

Environmental Influences The sound level meter should, ideally, be completely unaffected by environmental changes, other than the sound level. The standard specifies tests to check the effect on the sound level meter of variations in atmospheric pressure, mechanical vibration, magnetic and electrostatic fields, temperature and humidity.

Overall accuracy It is impossible to give a precise overall figure for the accuracy of each type of sound level meter. This is because the accuracy depends on a range of factors including the frequency content of the sound being measured and its direction relative to the microphone. A given sound level meter will therefore give a more accurate reading when measuring a beam of a "middle frequency" sound approaching the microphone at 0 degrees incidence than when measuring very high frequency sound approaching at large incidence angles. Nevertheless it is possible to give some rough guide to the accuracy of meters of different type when used under typically normal conditions, as follows:

± 0,7 dB	for type 0 meters
± 1 dB	for type 1 meters
± 2 dB	for type 2 meters
± 3 dB	for type 3 meters

Integrating sound level meters The performance requirements of integrating sound level meters are specified by BS EN 60804:1994. The standard is consistent with the requirements of BS EN 60651, for non-integrating meters, but, in addition specifies criteria and test methods for:

- integrating and averaging characteristics
- indicator characteristics
- overload sensing and indicating characteristics.

4.6 CALIBRATORS

The most important accessory of all, usually supplied by the manufacturer for each type of sound level meter, is an acoustic calibrator (**Figure 4.9**). This is a small portable battery operated device which when located precisely over the microphone capsule generates an accurately known sound level at the microphone. The calibrator is used to check that the instrument is working properly and also to detect any small day to day changes in sensitivity of the instrument. Any such changes may be corrected.

Most calibrators supply a calibration signal at only one frequency (e.g., 1 kHz) and one sound level (e.g., 114,0 dB), but there are some models which provide calibration at several levels and several frequencies.

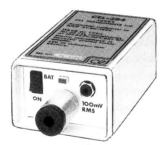

Figure 4.9 - CEL-284 acoustic calibrator [Casella]

Sensitivity changes may arise from changes in the environment, e.g., temperature, or in the circuit components, in the battery or in the microphone itself. The sound level meter should be calibrated at the beginning and at the end of each measurement, and at regular intervals in between if a prolonged measurement survey lasting several hours is being undertaken.

The acoustic calibrator is designed so that its output should remain accurately constant provided the battery voltage is adequate, but as with all sound measuring instrumentation it should be checked and calibrated itself, either by the manufacturer or by a calibration agency at regular intervals.

Internal Calibration Some sound level meters are also provided with an alternative, internal, calibration facility. This consists of a very stable signal generator which produces a signal of constant magnitude which can be switched into the sound level meter circuit, instead of the microphone signal. Based on the known sensitivity of the microphone the expected indication on the display device (e.g., the meter) can be checked, and any deviation corrected if necessary. This method of calibration cannot of course detect or remedy any changes in the microphone sensitivity, and for this reason acoustic calibration is much preferred, whenever possible.

4.7 MICROPHONES

Microphones used for accurate sound measurement need to be sensitive, stable, have a good frequency response and be able to operate over a very wide range of sound levels. There are basically three types of microphone, of which the first two satisfy the requirements in the previous sentence:

- Condenser microphones
- Electret microphones (pre-polarised)
- Piezo-electric microphones

The electret microphone is an adapted form of condenser microphone. The piezo-electric microphone, however, is only used on less accurate equipment. **Figure 4.10** shows an exploded view of a condenser microphone.

The microphone, whilst being the most important part of a sound level meter, is also the most delicate and easily damaged part; it is expensive to replace, and so needs to be handled with great care. The microphone determines the overall accuracy of the entire measuring system, and in some cases a sound level meter may be upgraded, say, from type 2 to type 1 accuracy by changing the microphone.

Pre-amplifier stages. Condenser microphones need to be supplied with a polarising voltage (typically 200 volts dc) whereas electric microphones do not, and so different preamplifier stages are needed for the two types. Nevertheless, many sound level meters will accommodate both types provided the correct polarising voltage is selected.

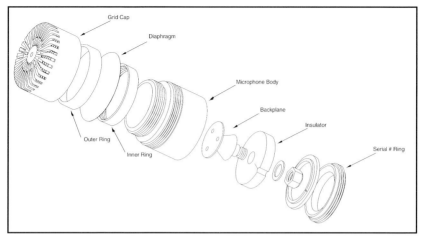

Figure 4.10 - An exploded view of a condenser microphone [Larson Davies]

Environmental considerations. Condenser microphones are sensitive, in general, to damage when the sound level meter is operated in wet and humid environments. This is because of the possibility that the polarisation voltage across the very small gap (of a few micron) between the metal diaphragm and backing plates of the condenser will cause arcing and sparking if condensation occurs. The design in **Figure 4.10** particularly takes humidity and temperature variations into account by providing a high leakage resistance through a proprietary quartz coating. Electret microphones do not suffer from this problem because a pre-polarised polymer film between the plates means that a polarising voltage is unnecessary.

Microphone frequency response A good quality microphone will have a flat frequency response and will have the same sensitivity (say, ± 2 dB) for different frequencies of sound over a range from about 20 Hz to an upper limit in excess of 10 kHz. The high frequency response will depend on the size of the microphone and on the type of sound field for which it is calibrated. (The smaller the diameter the greater the frequency response.) Half inch (12,5 mm) diameter microphones have a better high frequency performance than the one inch (25,4 mm) types which are gradually falling into disuse in modern instruments, despite their increased sensitivity. (The smaller the diameter the lower is the sensitivity.)

Quarter inch (6 mm) diameter microphones would allow measurements to even higher, ultrasonic frequencies. Many microphones are calibrated for freefield use and designed to be pointed at the source of sound so that the sound waves strike the diaphragm at 0 °, or normal incidence. Pressure response microphones are designed to be used at 90 °, or grazing incidence. When the sound field is diffuse, e.g., in reverberant situations the ideal microphone is one which has been calibrated for random sound incidence. The manufacturer's handbook will give guidance as to which type has been fitted and how it should be used.

Table 4.1 shows a selection of condenser microphones covering the three types of directivity (see section 3.5.3 a) and three diameters with their various sensitivities and frequency responses.

Diameter	6 mm		12,5 mm		25,4 mm	
Direction type (calibrated)	Free field	Random incidence / Pressure response	Free field	Random incidence / Pressure response	Free field	Random incidence / Pressure response
Sensitivity mV/Pa	3,7	1,3	14	11,7	50,7	45,5
Frequency response ±2 dB	4 Hz - 100 kHz	4 Hz - 80 kHz	4 Hz - 40 kHz	4 Hz - 21 kHz	2,6 Hz - 18 kHz	2,6 Hz - 8 kHz

Table 4.1 - Some condenser microphone sensitivities and frequency responses [based on Larson Davis microphones]

Cost. Overall, the condenser and electret types are comparable in cost and performance.

4.8 DATA ACQUISITION

Although a graph, chart or table gives a permanent record of the analysis of a noise, it is much better to have the measured data stored in some form useful for later analysis, e.g., an analogue or digital tape recording or solid-state disc storage. Such portable recordings could be necessary, for example, either because more sophisticated equipment is not available on site or maybe because only limited time was available, e.g., if machines are only operating in bursts of short duration.

It should be mentioned, also, that many sound level meters will have an alternative input, other than the microphone, to enable stored recordings to be analysed. It will then be possible to determine any noise measurement parameter by replaying the noise signal through the sound level meter.

Data acquisition in recent years has expanded in complexity and ability to record vast amounts of data in a small space and at a rapid rate. There is still room for the use of the older analogue tape recordings, but, more recently, the improved accuracy achieved with digital recordings has made DAT and solid state data acquisition the more acceptable techniques.

4.8.1 Tape recorders

Tape recorders are useful to

1. enable a subjective impression of the noise to be conveyed to a third party, for example, to demonstrate subjectively the effect of a noise control measure
2. provide evidence of the identity of the source of a particular noise and its time of occurrence, e.g., in cases of disputes about the onset of bursts of noise late at night
3. capture rare or unique transient events for later analysis, e.g., a controlled explosion or a demolition.

The tape recorder must have a sufficient dynamic range and an adequate frequency response to enable it to store and reproduce the signal to the accuracy required. Tape recorders with automatic level control and noise reduction circuits are not suitable for recording noise for measurement and analysis unless these facilities can be switched off.

Most tape recorders have a variety of different tape speeds. Whilst a slower speed will enable a longer period of time to be recorded, it must be remembered that the acceptable upper frequency will be reduced.

The use of multiple track tape recorders allows simultaneous recordings of two or more signals, e.g., of noise at two different locations, or of both noise and vibration from a machine, or of vibration in three directions (x, y and z). Comparisons between the different signals can help in the location of noise sources and the identification of transmission paths.

It is very important that all tape recordings are calibrated by the inclusion, on the tape, of a signal of known level (e.g., from an acoustic calibrator attached to the microphone). The details of all relevant instrument settings (such as the attenuator range controls on the sound level meter) and any changes to these settings during the recording must be carefully noted. Other details relevant to the measurement situation (e.g., date, time, microphone position, details of noise source, etc.) must also be noted. Ideally all these details should be noted both on the tape itself, i.e., via the spoken voice, and also on documentation attached to the tape.

4.8.2 DAT recorders

Digital Audio Tape recorders (DATs) are available with a better frequency response and dynamic range than analogue tape recording. They are suitable for
recording output from sound level meters or directly from a microphone via a
suitable preamplifier. These instruments have built-in ADCs and the acoustic signal is recorded on the tape in coded digital form. They have good frequency response over the audio range and can be adapted to operate down to dc, and have
a very high dynamic range, of about 80 dB. Digital to Analogue Converters (DACs)
enable the signal to be replayed through headphones or loudspeakers.

The very high dynamic range means that it is possible to record the entire
range of machine or environmental noise likely to be encountered in most situations without the need to adjust recording levels or sound level meter ranges. This
allows Environmental Health Officers and noise consultants to use DATs in the
investigation of noise complaints. Often the noise being complained of occurs
only intermittently, or during night-time periods when it is difficult for the investigator to be present to witness or measure the noise. A combination of DAT recorder and sound level meter, secured in a tamper-proof box is set up and calibrated and left in the complainant's home so that he or she can switch on the
instrument when the offending noise occurs. The tape recording can be used by
the investigator in a number of ways. It can help to verify the claims of the complainant about the noise, and be used in negotiations with the author of the noise
to help secure its reduction.

The use of DATs in machine noise analysis also enables a consistent trace to
be recorded whether the later analysis is to be of a comparative or a trend type.
Figure 4.11 shows an example of a DAT recorder with two tracks; in this case the
unit combines the functions of a sound level meter, a spectrum analyser and a
DAT recorder in a single device.

Figure 4.11 - DA-P1 DAT recorder & DF-1 sound level meter and analyser
[TEAC Corp]

4.8.3 Data loggers and environmental noise analysers

The level of many environmental noises varies widely with time and statistically based measurements have to be taken over long periods of up to 24 hours, or even longer. Examples are road traffic noise, train and aircraft noise, noise from building sites and from factories and commercial premises, and general community noise and noise from entertainment events.

Data loggers are devices which measure and store selected noise level parameters, e.g., L_{A90} L_{Aeq} L_{Amax}, etc. over selected time periods (e.g., every minute, or every 5 minutes, or every hour, etc.), for subsequent downloading into PCs, to allow presentation in graphical tabular or spreadsheet form. A microphone, connected to the data logger via a cable is left at the monitoring site over the required period (hours, days, weeks or months). Some hand-held sound level meters can also be used as data loggers.

Environmental Noise Analysers are, in effect, specialised data loggers which have been programmed to measure certain noise indices over a variety of pre-set measurement intervals. They often include their own printers and graph plotters.

Event Recorders In some situations, particularly at night time, noise levels are very low for most of the time and it is only occasional bursts of noise, which may be causing a disturbance, which are of interest. The analyser may be programmed to sample and analyse only noise events over a certain threshold level and to provide, at the end of the monitoring period, a list of such events, with their times of occurrence, durations and values of the selected noise indices, measured for each event. It is also possible for a tape recorder to be switched on automatically to record each event (and then switched off again afterwards) so that the events can be listened to and identified later. The same effect can now be achieved using digital signal processing.

4.8.4 Computers, data loggers and the virtual sound level meter

The use of digital signal processing in sound level meters has led to the increasing interfacing of sound level meters with computers so that signals from the sound level meter can be transferred, or 'down-loaded' to a computer, either for storage and continuing display, i.e., for data-logging, or for further analysis, or 'post-processing'. This has the advantage that noise level data from the sound level meter can be processed by databases, spreadsheets and graphical packages in the computer and very easily incorporated into reports, in the form of tables and graphs.

A natural extension of this process is to build the signal processing usually contained in the sound level meter directly into the computer, so that a signal from a microphone may be fed into the computer after only the minimum of signal conditioning, consisting of pre-amplification and some amplification. The processing and display of the microphone signal is then carried out in the computer, with the advantage of great flexibility, because different types of analysis, which may have required different dedicated instruments, such as for building acoustics, or environmental noise, may all be carried out within the same computer, simply by changing the software. Thus apart from the microphone and amplifier stages, all of the sound level meter is contained within the software of the computer, hence the term 'virtual sound level meter'. The increasing use and availability, data storage capacity and processing power of small portable 'laptop' PCs, which can easily be taken to the measurement site means that the trend towards use of PC based measurement systems will probably increase, but equally the use of dedicated hand-held sound level meters will probably also continue for certain applications.

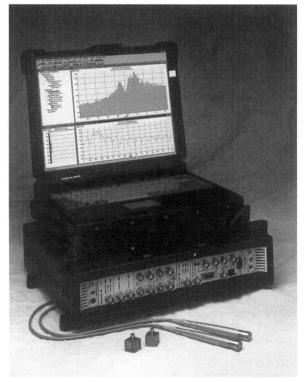

Figure 4.12 - The PULSE™ multi-analyser system [Brüel & Kjær]

With a suitable interface it is possible to use the vastly increased capacity of a Personal Computer (PC) to record and analyse the noise signals. For instance, the device shown in **Figures 4.12** is designed specifically for multi-channel and for frequencies from zero to 25,6 kHz, the period of time being limited solely by the hard disc of the PC.

4.9 OTHER INSTRUMENTATION

4.9.1 Real time analysers

The analysis refers to frequency analysis: usually octave or third octave bands. Real time, ideally, means instantaneously, or, since this is impossible, then fast enough for significant changes in the frequency spectrum to be followed as they happen. If the source of sound is the human voice, a fast moving vehicle or train, or a machine cutting timber, for example, the analysis must be completed and displayed in a small fraction of a second.

If the analysis is performed using analogue filters, then a bank of such filters is required, one for each frequency band. If the analysis is performed sequentially, i.e., switching from one filter to the next, throughout the range then the measurement cannot be performed in real time, since the noise spectrum may have changed during the analysis. Real time analysis may be achieved if the bank of filters is arranged in parallel, each with its own detector and indicator circuit, leading to a 'bar-chart' type of display.

In instruments using digital signal processing so-called 'digital filters' are used, in which the electronic components of the analogue filter are replaced by a set of software instructions, which can of course be changed easily and rapidly. Analysis performed using digital filters is usually considered to be performed in 'real time'.

Until the advent of the latest digital technology real time analysers were fairly bulky, 'stand alone' instruments, but real time analysis is now available in some digital hand held sound level meters.

4.9.2 Narrow band frequency analysers (including FFT analysers)

Although one sixth, one twelfth and one twenty fourth octave band analysers are available, the most commonly used form of narrow band analyser is the Fast Fourier Transform (FFT) analyser. These are similar to those widely used by condition monitoring engineers for the analysis of vibration signals, (as described in the handbook on *Vibration* in this series) but with the signal from the microphone replacing that from the accelerometer. They are used when it is required to identify the exact frequency as well as the amplitude of tonal components in the noise signal produced by a machine.

In the case of rotating machinery, for example, the frequency may be related to the rotational speed, and to the number of blades on a fan or turbine, or to the number of teeth on a gear, or the number of electrical poles on a motor or an alternator. From the variation in the level of the different harmonics of the rotational speed of the machine it is sometimes possible to identify the onset of failure and to identify possible causes such as bearing failure, out of balance or looseness of components.

FFT analysers (see **Figures 4.12 & 4.13**) work by capturing a sample of the analogue signal, digitising it and then performing the FFT process on the digitised sample, displaying the spectrum, and then repeating the process, i.e. capturing the next sample, and so on. For any given type of analyser the number of points or lines that make up the spectrum is fixed. The duration of each captured analogue sample, which relates to its statistical validity, the line spacing, which relates to the frequency resolution of the analysis, and the upper frequency limit of the spectrum are three inter-related parameters under the control of the user.

Figure 4.13 - Hand-held real-time analyser with FFT [Larson Davis]

Most FFT analysers have the ability to store and average repeated analyses to improve the statistical validity of the analysis, if the individual sample durations are very short, and to 'zoom in' on particular frequency ranges in the spectrum. Initiation of sample collection may performed manually or synchronised to some aspect of the noise producing mechanism, e.g., to the timing of an engine, or triggered by the advent of a transient event. Two channel FFT analysers allow correlation of noise and vibration signals, or from different parts of the same machine, or from noise arriving at the microphone from different transmission paths.

FFT analysis is now available in some all-digital sound level meters **Figure 4.13**, as well as in stand alone analysers.

4.9.3 Building acoustics analysers and measurement of machinery noise emission

Although sound insulation and reverberation time measurements can be carried out using hand held sound level meters the requirement that noise levels have to be the average of measurements taken at several different positions in a room, in two adjacent rooms (in the case of an airborne sound insulation test) and for a range of different frequency bands means that this would be time consuming and would also require a great deal of calculation. Specialist microprocessor controlled analysers are available which can measure store and process sound level data gathered from two rooms simultaneously and which produce an enormous saving in measurement time and analysis.

Sound power levels of machines may be measured with a similar range of equipment used for building acoustics. Measurements of sound pressure levels in either an anechoic or reverberant room have to be taken at different positions, and averaged, and then modified or corrected in some way. Reverberation time measurement may also be required for the reverberant room method. All measurements are required in octave or third octave bands.

4.9.4 Sound intensity meters

These are specialist devices which can detect and indicate the direction of flow of acoustic energy as well as measuring the magnitudes of sound intensity and sound pressure levels. The usual type has a probe which contains two identical microphones very close together. They can be used to locate sources within machines, identify weak links and flanking paths in sound insulation of partitions and enclosures, and to measure sound power levels of machines 'in-situ' and without the need for a specialist test environment of an anechoic or reverberant room.

4.9.5 Noise dosemeters, or 'dosimeters'

These are small portable devices, also called personal sound exposure meters, that can fit into a lapel pocket and be carried about throughout the day by the wearer to measure his/her noise exposure (**Figure 4.14**).

Figure 4.14 - Two dosimeters [Quest and Metrosonics]

A microphone connected to the instrument via a short flexible cable can be attached to the wearer's lapel, or to a hard hat if he/she is wearing one. The basic types measure only the accumulated noise exposure dose since the start of the measurement period. More sophisticated devices, often called data logging dosemeters, can display other parameters such as instantaneous sound pressure level, SEL and statistical Ln values. In addition they may be attached to a computer at the end of the working shift and the data can be down-loaded and then processed to display a minute by minute profile of the noise exposure pattern throughout the shift.

The more sophisticated types can also be used as sound level meters satisfying the requirements of type 1 or 2 of BS EN 60651:1994 depending mainly on the quality of microphone used. However, when worn as dosemeters they can only ever achieve type 2 performance because of effects of reflection and scattering of sound from the wearer's head and body. The performance requirements of dosemeters is described in BS EN 61252:1997.

4.9.6 Outdoor sound level meters and remote sound level monitors

Outdoor microphone kits are available for all weather noise measurements, including modified microphones with windshields which also provide rain cover, dehumidifiers and bird repelling spikes.

Output signals from remote noise measuring instruments may be transmitted via a modem linked to a telephone line, or via a radio transmitting modem, to a computer in a monitoring centre or control room. A more complex transmission is used in thermal power plants where the leakage flow of pressurised fluid through a hole or crack generates noise; in the case of the French 'Phonafuite' (English *leakage noise*), from Stell Diagnostic, the microphone uses a waveguide, generally 0,5 m long fitted with cooling fins allowing the microphone to operate at an acceptable temperature, and a 90 ° elbow is fitted to avoid direct radiation from the combustion chamber.

Outside noise monitors are also used for the monitoring for leakage of water **Figure 4.15**. In this case, to enable the microphone to identify noises below ground, it is shielded from the environmental and air noise by a covering. Both a meter and headphones are used to detect changes in noise initially. When a suspect region is identified a full analysis is undertaken - see the *Coxmoor Series* book on *Level, Leakage and Flow*.

Figure 4.15 - Leakage measurement by noise monitoring [Hermann Sewerin]

4.10 SELECTION OF SOUND MEASURING EQUIPMENT

The earlier sections in this chapter have indicated that, apart from the normal resolution and sensitivity, the main technical factors to be considered in a meter are:

- the type of analysis that the meter will undertake (e.g., SPL only, L_{eq}, Peak, Octave Bands, etc.),
- the meter's time weighting (i.e., slow, fast, impulse, peak)
- the range of the meter (e.g., 25 - 140 dB, 10 Hz - 20 kHz, etc.)
- how accurate it is (e.g., type 1 or type 2),
- the output from the meter (e.g., RS 232, ac, dc, etc.).

Other factors are:

- portability,
- how easy it is to use,
- the type and clarity of display,
- ruggedness,
- battery life (this is important if the meter is to be used for several hours at a time),
- service arrangements,
- flexibility in terms of availability of other measurement parameters, linking to other equipment, such as, a PC, and of possible upgrade at a later date and
- the cost.

CHAPTER FIVE

APPLICATIONS
AND
CASE STUDIES

5. Applications and Case Studies

5.1 APPLICATIONS

The machine and systems applications can be considered under three headings:

1. Machine/system condition
2. Monitoring to a standard
3. System diagnostics

5.1.1 Machine/system condition

Any new mechanical machine or system may well be first analysed by the human operator using one of the five physical senses, in particular, the ears and touch. High levels of vibration may even been seen, but vibration is more commonly heard or felt, and, depending on the experience of the operator, a rogue machine can be identified at this stage.

The next stage will be a desire to quantify the sound in some way using acoustical techniques as described in this book. Human experience is subjective, and a more precise and repeatable technique is needed so that even inexperienced personnel can achieve good results. Having determined that noise levels are exceptionally high, a more permanent system may be installed which uses fixed vibration sensors (such as accelerometers) or alternative condition monitors.

Noise is thus seen as an important part in the developing monitoring programme. This is apparent in the case studies given in section 5.2 where most involve the early assessment of a situation - the development stages - rather than the final monitoring technique used in the field.

The noise monitoring, however, may play an even greater part in certain applications where vibration signals, emanating in noise, are less confused by the environment. These can be analysed more quickly and more distinctly. Fans are one example of this kind of application. Noise monitoring is also extremely valuable where a large number of quite small systems or components need to be assessed rapidly and there is insufficient time for a fixed sensor to be attached - see later in the Case Studies (section 5.2).

5.1.2 Monitoring to a standard

Applications here may be more associated with the human being, in the sense of noise nuisance and disturbance. This is essential monitoring in those applications where noise is considered a health hazard. Some noise is always going to be generated by moving components, even on a macro scale, but when the design is faulty or the build has developed a flaw, then the noise levels may well rise to unacceptable proportions.

Various standards associated with governing bodies, national and international, are described in the Standards section in Chapter 7. There are also local standards which relate to two-party agreements, for instance, and for these the noise levels need to be measured and acceptable levels decided by both parties; measurement would then be applied in a consistent manner at certain times and under certain conditions

5.1.3 System diagnostics

The path taken by sound waves is rarely singular. In other words, there may be many routes, both in solid structures and in air; but the analysis of the noise levels over a variety of positions will indicate the primary paths. This, like the previous section, is mainly concerned with human reaction but it does also have an influence on the machinery itself. For instance, noise pressure waves can affect delicate instrumentation.

Another field in this application is that of military stealth, or the effect on wild life where machinery is used in its vicinity.

5.2 CASE STUDIES

Numerous case studies exist where an operator has sensed from his own hearing that a machine is malfunctioning. In most cases, these imply that the machine has reached a fatal condition, such as the driver of a car hearing a big-end rattle well after the bearing has failed. More skilled operators have detected earlier signs and have taken corrective action at the most appropriate time, but, nevertheless, this has been very subjective although useful.

This chapter does not need to list such subjective cases - they are well-known - instead it is devoted solely to conditions where an instrument has detected a change through an air-borne noise generation. The change may have been detected by other means at the same time, but the noise sensing initially played a major part.

The section is separated into three parts

- *experimental testing*, where it was known that a fault existed - it was included as part of the testing process - but noise analysis was used to singularly identify the fault,
- *production testing*, where components are checked for acceptability from the noise generated under test, and
- *industrial or field use,* where it was thought that a fault might be present, or could develop, and noise would, and indeed did, indicate its presence.

5.2.1 Experimental testing

a. Automobile gear box - faulty bearings, damaged gear teeth

[University of the West of England, Bristol, **Johnson (1993)**]

A five-speed constant-mesh gearbox of a Ford Sierra design was put under investigation to determine whether an automatic acoustic analysis could replace the subjective assessment of production inspectors. Basically the inspectors, after some experience, were able to detect certain levels of fault relating to misalignment, bearing problems and damaged teeth.

Although no testing was able to proceed to the industrial environment, three separate gearboxes were examined in laboratory conditions: -

1. A good gearbox
2. One containing faulty bearings
3. One containing a gear wheel with damaged teeth

Tests were undertaken at constant load, and the whole arrangement was fitted into a sound reducing enclosure. The microphones was positioned 1 m in front of the gearbox and 1 m above the floor, and attached to a Brüel & Kjær type 4433 sound intensity analyser.

Figure 5.1 shows the range of sound pressure levels detected across the five gear ratios over a range of drive speed for each of the three gearboxes. This is the rms sound pressure level which easily detected the difference between the good boxes and the poor ones.

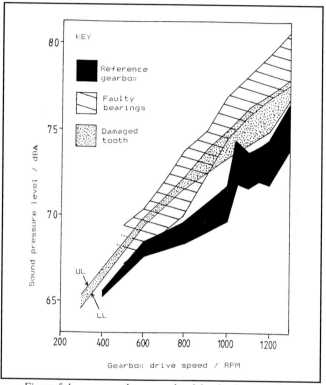

Figure 5.1 - rms sound pressure level for the three gearboxes

Analysis was also undertaken using the time domain with just the first gear engaged - the first gear having the faulty gear wheel. Examples of the quite different results are shown in **Figure 5.2**.

Several different analyses were used both with the as-supplied gear boxes and with those with components swapped over. These used frequency analysis, Kurtosis, Crest, Cepstrum analysis and three special methods called 'power deviation', 'ratio 1' and 'CROS'. All of these were able to detect the differences to a greater or lesser degree. The CROS (correlated rms in overlapping sections) parameter proved to be the most reliable at separating the acceptable from the non-acceptable gears.

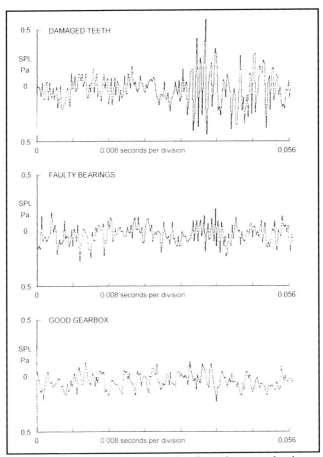

Figure 5.2 - Time domain results of sound pressure level

The CROS value is determined using an examination of the rms value of small time periods which overlap. **Figure 5.3** shows how this is undertaken. The equation used is:-

$$CROS = \frac{n\Sigma xy - \Sigma x \Sigma y}{\sqrt{(n\Sigma x^2 - (\Sigma x)^2)(n\Sigma y^2 - (\Sigma y)^2)}}$$

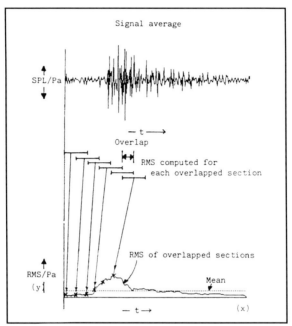

Figure 5.3 - The procedure for the correlated rms value in overlapping sections of time
Johnson (1993)

b. Fibre-optic gyroscopes - light source and fibre failures

[Reference **Dankwort (1998)** WO Patent 98/54544 associated with Honeywell Inc, Minneapolis, Dec. 1998]

Fibre-optic gyroscopes are optical inertia rotation sensors which use the Sagnac effect to detect rotation about an axis. (The Sagnac effect essentially involves the detection of the phase-shift in light waves.) There are several faults which can develop in such devices and some of these cause a change in the noise generated.

The noise output is roughly proportional to light power, and the dimming of the light transmitted would normally reduce the noise level. Such essential features, therefore, as the light source, the fibre splicing and couplers, as well as the optical-phase modulator and photo-detector assembly have all been found to change the noise level when a fault developed.

Although this is not yet available for industrial use the testing has indicated that it will be an invaluable means of monitoring such delicate and critical units which use fibre-optics.

5.2.2 Production testing

a. Production line testing of vehicle engine dampers [Pro-Test]

[Peter Wilson, INVC]

Lanchester dampers are used to reduce the torsional vibration on engines. In this example the prefilled dampers were able to be checked acoustically for the level of filling before despatch for engine build.

No additional production process was required, but a microphone was placed close to the serial number engraving printer. This printer generated sufficient impact sound which was modified according to the quantity of viscous fluid within the damper.

Pro-Test software was used to examine the impact response and automatically provide visual and audible warning when the filling was inadequate, and stop the acceptability of the product. Pro-Test is a virtual instrumentation test system suitable for a wide range of applications, such as, gear mesh knock, sunroof noise, glass breakage, crack detection, etc.; it automatically generates a pass/fail signal by comparing noise signatures with stored criteria. **Figure 5.4** shows the variation in signal between a fully filled damper and one which was partially filled or empty. The frequency range 0 to 20 kHz was used.

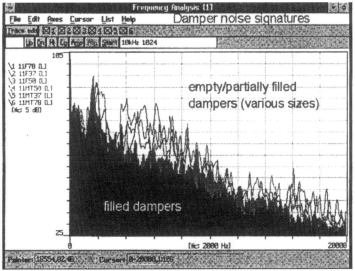

Figure 5.4 - Noise signatures for filled and empty/partially filled dampers [INVC]

b. Quality testing of transfer box whine

[Reference: **Plant (1998)** associated with Diagnostic Instruments and Land Rover]

One of the inherent problems of land vehicles is the whine emanating from the drive. This is not only a nuisance to personnel, it also demonstrates that a poor distribution of load is occurring, probably due to wear or poor fit, or even poor design.

The objective of the programme instituted on the Land Rover Discovery was that of identifying and resolving the whine in the transfer box - the unit which distributes the load from the 4x4 gearbox to the axles. This had been a recurring problem at 35 mph in some of the units manufactured, but not all. It was decided to use the Diagnostic Instruments LineCHECK, with a microphone, as a portable within the cab during road testing (**Figure 5.5**).

Figure 5.5 - The LineCHECK within a Land Rover [Diagnostic Instruments]

The signature, or spectrum, detected for acceptable noise levels was compared with those experienced in use. The driver/operator was responsible initially for indicating when the noise reached an objectionable level, i.e. when the whine became prominent. At that stage, the LineCHECK was prompted by the touch of a button so that it could note which spectra would be unacceptable in future tests.

Over a period of six weeks Land Rover was able to identify the reason for high levels of whine - from the spectra of the units which caused the whine - and make appropriate modifications (to the power train) so that the acceptability was more consistent. The axles causing the fault had to finally be checked individually with accelerometers to identify the source and speeds more precisely.

c. Bearing test machine

[Reference **Drives & Controls (1998)** associated with RHP Bearings]

The N2 noise tester is an automatic test process for assessing the quality of small precision bearings under load. In particular, it looks at the noise spectrum generated at 1800 rev/min.

Individual bearings are fed, every three seconds, onto a spinning precision air-spindle. This is driven by a specially designed low-noise belt, controlled within ± 2% of the 1800 rev/min. A pre-set load is applied to one side of the bearing and microphone probe measures the noise signal. A second test applies the load on the opposite side of the bearing to check on the opposite side.

Four frequency bands are analysed and if a bearing exceeds certain limits in any of the bands, or a combination of bands, then the bearing is rejected.

d. Imperfections in steel rods

[Diagnostic Instruments Ltd. **Diagnostic Instruments (1996)**]

Shearing steel rod into suitable short component lengths at typically one cut per second can cause imperfections in the lengths. These flaws are normally indiscernible until the components have received further work, and may, even, remain to be built into the final product.

It is not surprising, though, that a flawed component makes a different sound to one which is homogeneous. The problem, however, has been to detect such small changes. What has been found is that the imperfect component yields a cracking sound 10 ms before the bang produced by the cutting tool in the same action. A subtle inspection system has been developed that uses both an accelerometer (on the cutting tool) and a microphone to analyse the noise and vibration together.

The system triggers an alarm if the sound differs from the correct signature, and the cropping machine is immediately shut down. Early trials have shown a 100% success rate.

e. Fault diagnosis system for motor vehicle engines

[Daifuku Co., Ltd, Osaka, Patent DE 198 22 908 -**Akishita, Li & Kato (1998)**]

In order to test for the correct dimensional tolerances in manufacture, a test arrangement for an IC engine is used as shown in **Figure 5.6**. The engine is motor driven through a reduction gearbox and includes a flywheel before the final drive of the engine. A microphone sensor is used above the engine, and various displacement and rotational sensors are fitted..

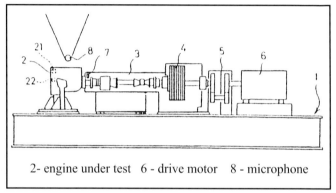

2- engine under test   6 - drive motor   8 - microphone

Figure 5.6 - Engine test centre [Daifuku]

Although no combustion occurs, and hence the forces are not as in final use, the repeatability of the test arrangement enables errors, in such components as the tappets and shafts, to be detected before the engine is assembled into a motor vehicle.

5.2.3 Industrial or field use

a. Automobile body seals - in car monitoring

[Ford Motor Company, CEL Instruments, **Eureka (1998)**]

Car noise is primarily tested as regards nuisance to the passengers (and, maybe, also pedestrians). However, the source of such noise may be due to faulty fitting or manufacture of components and these need to be identified before the vehicle is sold to the public. One application which has proved successful is that of window seal faults.

In the traditional dBA approach, a small change in a window seal will produce a measurement that differs by a few tenths of a dB - which is less than the variation from test to test. However, the loudness technique developed by CEL results in a difference of several Sones - a repeatable and objective change beyond the normal variations of testing. This is based on the ISO 532 Part B standard.

Figure 5.7 - The 593 real-time sound level analyser [CEL]

The CEL 593 real-time sound analyser - **Figure 5.7** - permits loudness measurements to be made real-time in field conditions. 10 ms of 1/3 octave spectral data is gathered and either a 100 or 500 ms loudness level computed. A time history is built up along with the 1/3 octave spectra used to compute the loudness.

Early results at the Ford Motor Company have been satisfactory and it is expected savings of more than $250 000 per year at a single plant could be achieved. This is around 20 times more than the cost of the monitor in the first year alone

b. Underground surface conveyors - bearing idlers

[Vipac Engineers & Scientists, New South Wales, Australia, **Vipac (1996)**]

Underground conveyors for coal mines are notoriously prone to failure of the bearing idlers. They are numerous - of the order of 5000 per kilometre - and have on average a life span of five years which can lead to up to ten failures per day per kilometre. The importance of highlighting a failure in the early stages is shown by belt damage and fire risk due to seized idlers.

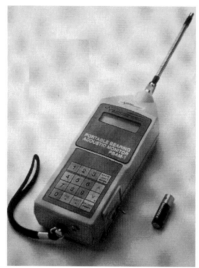

Figure 5.8 - The Bambino bearing acoustic monitor [VIPAC]

VIPAC successfully developed a portable bearing acoustic monitor PBAM-1 ('Bambino') - **Figure 5.8** - which could be vehicle mounted and measure the acoustic signals of the idlers at a distance of between 0,5 and 1,0 m. The incorporation of suitable software has enabled the Bambino to successfully extract a bearing fault signature even in a reverberant underground tunnel. An omnidirectional microphone continuously senses the ambient sound pressure level as the operator moves at walking speed along the conveyor route.

It has been found that, from a knowledge of the belt speed and roller diameter with the bearing type characteristics, the following features are detected:-

- bearing faults
- build-up on rollers
- ranking of severity

False alarms have been successfully isolated by identifying noise sources unrelated to the bearing faults and roller build-up.

c. Vehicle pneumatic tyres - rolling tests and road tests

[Dunlop SP Tyres Ltd and Brunel University Neural Applications Group - **Harris (1994)**]

The noise generated from automobile tyres, in operation on roads, is a good indication of their condition and of their ability to retain their performance on the type of road and speed under test.

Probably the first tyre noise tests were undertaken around 1847 - **N & V (2000)** - by Robert Thompson with the intention of determining the best tread for the available road surfaces. Road rolling tests, almost 150 years later, are designed to test durability and extreme condition performance of a tyre as it is run on a simulated road surface at very high speeds - even to the point of destruction. When removed from the test apparatus, the tyre is analysed to determine how it malfunctioned. From this it is possible to identify design improvements or faults in the quality control procedure.

This durability test has previously been supervised by a human operator, who detects the point where the tyre becomes faulty from characteristic changes in the sound of the tyre on the 'road'. It is important that the operator identifies this point exactly and ends the test before any further damage is done to the tyre - otherwise this will interfere with the analysis.

The approach is both expensive and very demanding on the operator who has to concentrate consistently over a long period. In addition, as each tread pat-

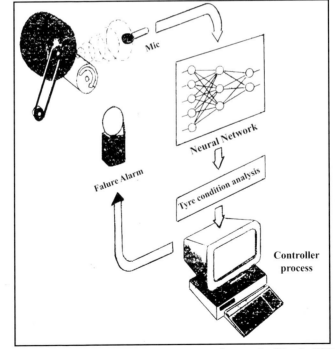

*Figure 5.9 - Tyre noise monitoring using neural networks [**Harris (1994)**]*

tern is different, and the road speeds cover a wide range, the actual sound harmonics and intensity produced vary considerably.

Dunlop SP Tyres UK. and Brunel University Neural Applications Group have developed a similar test but one which uses a learning neural network to determine the condition of tyres during the high-speed road tests. The system uses a microphone placed in the test room to pick up the road noise of the tyre. - **Figure 5.9**. The sound is then pre-processed on-line to produce a frequency spectrum that is used as the input data for the neural network.

The neural network has to determine an initially acceptable noise spectrum, which it does in the first few minutes of each test and after any speed changes.

Once trained, the neural network monitors the sound and raises an alarm as soon as it detects the change in road noises that indicate a fault - making it possible for the tyre to be removed before any further damage can occur.

The neural network was found to be as effective as a human expert in determining when a sample tyre was beginning to fail, and its use results in considerably reduced costs.

[Transport Research Laboratory **Abbott, Phillips & Nielson (2001)**]

A specific road test procedure, called TRITON, has been developed by the Transport Research Laboratory. The TRITON system uses a 7,5 tonne vehicle having a tyred wheel and microphones within a sound-proof enclosure so that a close-proximity noise test can be undertaken on real roads at a range of speeds.

In passing, it should be mentioned that TRL are also developing a feedback system on the noise of a vehicle. (It relates not only to the tyre/road interaction but also to the way a driver drives - aggressively or passively.) **Figure 5.10** shows the features involved on a typical car.

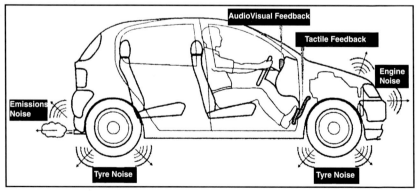

*Figure 5.10 - Driver feedback system on car noise [**Abbott et al (2001)**]*

d. Corona discharges in a transformer

[Fortum Service Oy, Finland **Kokko (2000)**]

Loose parts or strong partial discharges can cause such noise inside a transformer that acoustical waves are propagated inside the oil to the walls of the transformer. These waves can be detected by antenna sensors attached to the wall, and the failure point can be detected by the arrangement of several of these sensors (e.g., by measuring the arrival times of different sensors and determining the location by calculation).

Acoustical methods have also been successfully used to identify unwanted noise in insulators. By this method it is possible to detect corona-type discharges on the surface of the insulator.

e. Railway wheels and track-wheels

[Danish State Railways (DSB) - **Rasmussen (1998)**]

The reaction of wheel and track may produce a marked change in noise when a crack or wear is developing on the wheel or in the track. Once a defect has started to develop, the wheel condition will rapidly deteriorate and the cost of repair rises dramatically. Early detection is vital so that a minimum of around 0,1 mm may be skimmed off the surface of the wheel; (later detection may result in up to 5 mm being removed or even the scrapping of the wheel).

Two systems have been used, one with a microphone stationary alongside the track (for checking wheels), and the other a microphone carried on the train itself for checking both wheels and track (see the following case study)

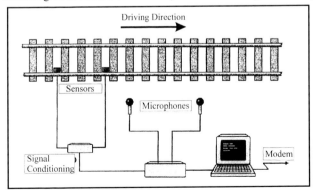

Figure 5.11 - Train wheel noise monitoring [Danish State Railways]

The arrangement used by the Danish State Railways (DSB) is shown in **Figure 5.11**; it is designed to check the condition of the wheels. The two positional sensors (at 10 m and 20 m from the first microphone) determine the speed of the train and trigger the microphone recording. The G.R.A.S type 41CM microphones (**Figure 5.12**) are rugged units specially designed for permanent outdoor use in harsh environments, and are place 0,7 m at the side of the track.

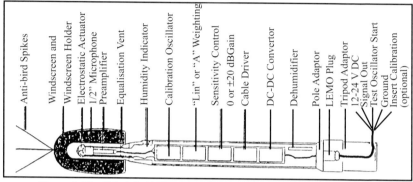

Figure 5.12 - G.R.A.S. 41CM microphone built for permanent outdoor use
[Rasmussen (1998)]

The microphone signals are digitised and filtered to remove the low frequency pulses associated with the air pressure at the front and back of the train. Individual bogie noises are easily identified and their noise level recorded with the train speed so that a noise index may be calculated independent of speed. **Figure 5.13** shows an example for one particular train over a period of time, a two-day maintenance where the wheels were machined and the improvement seen immediately thereafter.

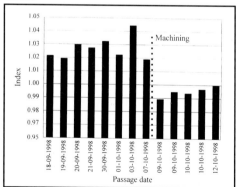

Figure 5.13 - Speed corrected noise before and after machining [DSB]

f. Railway wheels and track - track

[AEA Technology Rail BV, Netherlands - **Dings (2001)**]

Although this work is being undertaken for the benefit of passengers - as regards noise experienced within carriages - it also offers an ability to detect faults, or severe wear, in the tracks or wheels.

The difference between a good wheel/track noise and a 'rough' one can be as great as 20 dBA. In order to measure the track roughness accurately accelerometers may be used on the axle box or a microphone positioned between two wheels of a wheelset while the train is passing over the track. **Figure 5.14** shows the spectrum of results for four rails where instead of frequency the wavelength is shown, i.e. five centre-wavelengths of the octave bands. (It should be noted that the frequency f is given from f = v/l where v is the speed of the train and l is the wavelength.)

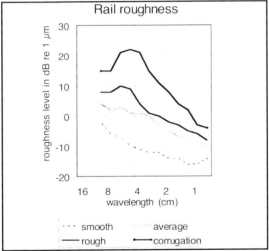

*Figure 5.14 - Rail roughness spectra measured from a microphone on a train [**Dings** (2001)]*

Whilst this is relatively early on in the life of this testing it has already come up with some surprising results. For instance, it was found that the smoothest rail was the oldest and the roughest the newest! It is likely that a certain amount of 'running in' will be experienced which would smooth the rails with time, but this will mean that any severe cracks or distortion of the rail will become more identifiable with time.

CHAPTER SIX

BUYERS' GUIDE

6. Buyers' Guide

6.1 COMPANIES

Manufacturers, suppliers and fitters of sensors and systems, consultants, publishers and training establishments for noise and acoustics monitoring

Academic Press Ltd - 14-28 Oval Road, London, NW1 7DX, UK
Tel: (+44) 207 482 2893   Web: www.accademicpress.com/jsv

Publishers of the *Journal of Sound & Vibration*

Acoustic Associates - 7A Flag Business Exchange, Vicarage Farm Road, Peterborough, PE1 5TX, UK
Tel: (+44) 1733 896346   Fax: (+44) 1733 567779
e-mail: Acoustic.Associates@btinternet.com

Noise and vibration consultancy, including measuring and identifying noise and vibration sources

Acoustical Society of America - 500 Sunnyside Blvd, Woodbury, NY 11797-2999, USA
Tel: (+1) 576 576 2360   Fax: (+1) 576 576 2377
e-mail: asa@aip.org

Society which includes the issue of a regular journal

AcSoft Ltd - 8B Wingbury Courtyard, Leighton Road, Wingrave, Aylesbury, HP22 4LW, UK
Tel: (+44) 1296 682686   Fax: (+44) 1296 682860
e-mail: jshelton@acsoft.co.uk

Equipment suppliers specialising in virtual sound level meters.
UK agent for 01dB

AEA Technology Rail BV - PO Box 8125, 3503 RC Utrecht, The Netherlands
Tel: (+31) 30 235 7073   Fax: (+31) 30 235 4163
e-mail: pieter.dings@nl.aeat.com

Consultancy and testing associated with railway acoustics.

Anthony Best Dynamics Ltd - Holt Road, Bradford on Avon, Wilts, BA15 1AJ,
UK
Tel: (+44) 1225 867575   Fax: (+44) 1225 864912
e-mail: info@abd.uk.com   Web: http://www.abd.uk.com

Drive-line noise and vibration test systems. Design and production of specialised
test systems. In-house testing.

W S Atkins, W.S., Noise & Vibration - Woodcote Grove, Ashley Road, Epsom,
Surrey, KT18 5BW, UK
Tel: (+44) 1372 726140   Fax: (+44) 1372 740055
e-mail: rwowen@wsatkins.co.uk   Web: http://www.wsanoise.com

Noise 98 software. Consulting engineers, surveys, assessments and machine monitoring.

Automated Analysis Ltd, Unit C, Enak House, Redkiln Way, Horsham, West Sus-
sex, RH13 5QN, UK
Tel: (+44) 1403 218718   Fax: (+44) 1403 218728
e-mail: automated@fastnet.co.uk   Web:

Computer software for complex noise problems

Brüel & Kjær [world headquarters] -DK-2850 Nærum, Denmark
Tel: (+45) 42 80 05 00   Fax: (+45) 42 80 14 05
e-mail:
UK Branch: Bedford House, Rutherford Close, Stevenage, Herts, SG1 2ND, UK
Tel: (+44) 1438 739100   Fax: (+44) 1438 739199
e-mail: info@bkgb.co.uk   Web: http://www.bk.dk

Manufacturers of microphones, advanced sound level meters and real-time analysers and PC interfaces.

Burgess Industrial Acoustics Ltd - 1 Canton House, Wheatfield Way, Hinckley, Leics, LE10 1YG, UK
Tel: (+44) 1455 619453   Fax: (+44) 1455 631942
e-mail: sales@burgess.ltd.uk   web: http://www.burgess.ltd.uk

Consultancy and full noise surveys. Enclosure design.

Cambell Associates - Chiswell Cottage, 11 Broad Street, Hatfield Broad Oak, Bishop's Stortford, Herts, CM22 7JD, UK
Tel: (+44) 1279 718898   Fax: (+44) 1279 718963
e-mail: info@cambell-associates.co.uk

Suppliers of equipment. Agents for Norsonic

Casella Group Ltd, Regent House, Wolseley Road, Kempston, Bedford, MK42 7BR, UK
Tel: (+44) 1234 844100   Fax: (+44) 1234 841490
e-mail: info@casella.co.uk   Web: http://www.casella.co.uk

Real-time sound level analysers carrying out simultaneous measurement of multiple time, frequency and amplitude weighted sound level data real-time octave and one-third octave band frequency options.

CEL Instruments Ltd. - see *Casella Group*

Castle Group Ltd - Salter Road, Cayton Low Road Industrial Estate, Scarborough, North Yorks, YO11 3UZ, UK
Tel: (+44) 1723 584250   Fax: (+44) 1723 583728
e-mail: sales@castlegroup.co.uk   web: www.castlegroup.co.uk

Manufacturer of noise dosimeters

Cirrus Research plc - Acoustic House, Bridlington Road, Hunmanby, North Yorks, YO14 0PH, UK
Tel: (+44) 1723 891655   Fax: (+44) 1723 891742
e-mail: sales@cirrusresearch.co.uk Web: http://www.cirrusresearch.co.uk

Sound level meters with data logging capacity including L_{eq}, Peak, I_{max}, I_{min}, SEL and three user defined L_ns. Digital integrated sound level meter. Permanent and portable monitoring systems.

Civil Engineering Dynamics - 83/87 Wallace Crescent, Carshalton, Surrey, SM5 3SU, UK
Tel: (+44) 20 8647 1908   Fax: (+44) 20 8395 1556
e-mail: Web: www. environmental.co.uk

Consultancy and hire

Cranfield Data Systems (CDS), - *Cranfield Aerospace Ltd*, Building 125, Wharley End, Cranfield, Bedford, MK43 0AL, UK
Tel: (+44) 1234 751214   Fax: (+44) 1234 751871
e-mail: sales@cds-international.co.uk

Software for noise analysis including PRISM database and SAPPHIRE multi-channel analysis package. Many applications are covered, including rotating machinery, noise route tracking, pass-by noise simulation, acoustic intensity, sound quality engineering and analysis of jet engine inlet distortion.

Diagnostic Instruments Ltd. - 2 Michaelson Square, Kirton Campus, Livingston, West Lothian, EH54 7DP, UK
Tel: (+44) 1506 470011   Fax: (+44) 1506 470012
e-mail: sales@diaginst.co.uk   Web: www.diaginst.co.uk

Manufacturer of PC-based analysers offering high performance, multi-channel, real-time spectrum analysis. Courses and seminars.

Gracey & Associates - High Street, Chelveston, Northants, NN9 6AS, UK
Tel: (+44) 1933 624212   Fax: (+44) 1933 624608
e-mail: hire@gracey.com   Web: www.gracey.com

Hire equipment including Norsonic

Hermann Sewerin GmbH, Postfach 2851, D-33326 Gütersich, Germany
Tel: (+49) 5241 934-209   Fax: (+49) 5241 934-444
e-mail: Sandra.Vogt@Sewerin.comWeb: http://www.sewerin.com

Electro-acoustic leak noise detector with microphone and head-phones, particularly suitable for water leakage.

Hodgson & Hodgson Group Ltd, Crown Business Park, Old Dalby, Melton Mowbray, Leics, LE14 3NQ, UK

Tel: (+44) 1664 821800   Fax: (+44) 1664 821830
e-mail: postmaster@noiseco.demon.co.uk Web: http://noiseco.demon.co.uk

Acoustic consultancy primarily with the noise problem solving objective, including also anechoic chambers. Noise Control Centre.

Industrial Noise and Vibration Centre - Burnham House, 267 Farnham Road, Slough, Berks, SL2 1HA, UK
Tel: (+44) 1753 570044   Fax: (+44) 1753 570311
e-mail: consult@invc.co.uk Web: http://www.invc.co.uk

An independent engineering consultancy specialising in the field of noise and vibration. Noise surveys and systematic control. PC based virtual instruments - hardware and software. Training.

Institute of Acoustics - 77A St. Peter's Street, St. Albans, Herts, UK
Tel: (+44) 1727 848195   Fax: (+44) 1727 850553
e-mail: ioa@ioa.org.com   Web: www.ioa.org.uk

Annual register of members including noise consultants and suppliers of acoustics and noise control products and materials. Courses at accredited centres, diplomas in acoustics and noise control.

Institute of Sound and Vibration Research - *University of Southampton,* Southampton, Hants, SO17 1BJ, UK
Tel: (+44) 1703 592162   Fax: (+44) 1703 592728
e-mail: m.c.lower@soton.ac.uk

Consultancy service with extensive laboratory facilities.

Larson Davis Inc., 1681 West 820 North, Provo, UT 84601, USA
Tel: (+1) 801 375 0177   Fax: (+1) 801 375 0182
e-mail: mktg@larsondavis.com   Web: http://www.larsondavis.com
UK distributor: *Proscon Environmental*

Dosimeters, sound level meters, octave band analysers and real time analysers

Metrosonics Inc - see *Quest Technologies* and *P.C.Werth*

Noise Control Centre - see *Hodgson & Hodgson Group*

Noise & Vibration Worldwide - Journal, c/o Multi-Science Publishing Co Ltd, 5 Wates Way, Brentwood, Essex, CM15 9TB, UK
Tel: (+44) 1277 224632   Fax: (+44) 1277 223453
e-mail: mscience@globalnet.co.uk Web: www.multi-science.co.uk

Publisher of monthly journal, etc.

Norsonic - P.O. Box 24, N-3408 Tranby, Norway
Tel: (+47) 32 85 20 80   Fax: (+47) 32 85 22 08
e-mail:

Manufacturers of sound level meters, sound analysers, real-time analysers, microphone equipment and calibrators

Ono Sokki Co Ltd. - Head Office: 1-16-1 Hakusan, Midori-ku, Yokohama 226, Japan
Tel:
e-mail:

Real-time 1/1 and 1/3 octave sound level meters and analysers

Palmer Environmental Ltd, Ty Coch House, Llantarnam Park Way, Cwmbran, Gwent, NP44 3AW, UK
Tel: (+44) 1633 489479   Fax: (+44) 1633 877857
e-mail: info@palmer.co.uk   Web: http://www.palmer.co.uk

Acoustic logging, listening stick, microphone, etc. for water leakage

P.C Werth Ltd - see *Werth, P.C. Ltd*

Proscon Environmental Ltd, Abbey Mill, Station Road, Bishops Waltham, Southampton, Hants, SO32 1GN, UK
Tel: (+44) 1489 891853   Fax: (+44) 1489 895488
e-mail: info@proscon.co.uk Web: http://www.proscon.com

UK supplier of *Larson Davis* equipment

Quest Technologies Inc. - 1060 Corporate Center Drive, Oconomowoc, WI 53066, USA
Tel: (+1) 414 567 9157   Fax: (+1) 414 567 4047

e-mail: Web: http://www.quest-technologies.com
UK distributor: *P.C.Werth*

Sound measurement and analysis equipment including level meters and noise dosimeters.

Recording Systems Ltd - Unit 8a, Acorn Business Centre, Cublington Road, Wing, Beds, LU7 0LB, UK
Tel: (+44) 1296 682626   Fax: (+44) 1296 688700
e-mail: sales@recordingsystems.co.uk

UK distributors for TEAC noise recording instrumentation

Society of Environmental Engineers - Owles Hall, Buntingford, Herts, SG9 9PL, UK
Tel: (+44) 1763 271209   Fax: (+44) 1763 273255
e-mail: see@owles.demon.co.uk   Web: www.environmental.org.uk

Professional society promoting awareness of environmental engineering. Services include journal and newsletter, test house users & buyers guide, vibration, shock and noise group, etc.

Spectral Dynamics (UK) Ltd - Fulling Mill, Fulling Mill Lane, Welwyn, Herts, AL6 9NP, UK
Tel: (+44) 1438 716626   Fax: (+44) 1438 716628
e-mail: spectral@sd-star.co.uk   Web: www.spectraldynamics.com

Manufacturers of vibration and sound analysis equipment

Stell Diagnostic - Instrumentation and Nuclear Systems Dept., 565 rue de Sans Souci, F-69760 Limonest, France
Tel: (+33) 4 72 20 91 00   Fax: (+33) 4 72 20 91 01

Suppliers of the *Phonafuite* for leak detection by microphone.

TEAC Corp - 3-7-3, Naka-cho, Musashino, Tokyo 180-8550, Japan
Tel: (+81) 422 52 5016   Fax: (+81) 422 52 1990
e-mail: Web: www.teac.co.jp

Digital acoustic system and DAT recording equipment (See *Recording Systems*)

Vipac Engineers and Scientists, Melbourne, Australia
Tel: (+61) 3 9647 9700   Fax: (+61) 3 9646 4370
e-mail:

Portable bearing acoustic monitor, consultancy and modelling, etc.

Werth, P.C. Ltd, Audiology House, 45 Nightingale Lane, London, SW12 8SP, UK
Tel: (+44) 20 8772 2700   Fax: (+44) 20 8772 2701
e-mail: pcwerth@pcwerth.co.uk   Web: http://www.pcwerth.co.uk

UK suppliers of *Metrosonics* and *Quest* equipment

W S Atkins Noise & Vibration - see *Atkins, W.S.*

6.2 PRODUCTS, SERVICES AND HIRE

Analysers

Brüel & Kjær
Casella Group
Diagnostic Instruments
Larson Davis
Norsonic
Ono-Sokki
Proscon
Spectral Dynamics
Quest Technologies

Calibrators

Brüel & Kjær
Casella
Cirrus
Norsonic

Calibrators

Beran Instruments

Data acquisition, recorders & loggers

Brüel & Kjær
Recording Systems
TEAC

Dosimeters

Brüel & Kjær
Campbell
Castle Group
Casella Group
Cirrus

Larson Davis
Metrosonics
Norsonic
Proscon
Quest Technologies
PC Werth

Hire

Civil Engineering Dynamics
Gracey Associates

Leak noise detection

Hermann Sewerin
Palmer Environmental
Stell Diagnostic

Microphones and accessories

Brüel & Kjær
Hermann Sewerin
Larson Davis
Palmer Environmental

Monitoring systems

Anthony Best Dynamics
WS Atkins Noise & Vibration
Casella
Cirrus Research
INVC
Vipac Engineers & Scientists

Software

AcSoft

W.S. Atkins Noise & Vibration
Automated Analysis
Brüel & Kjær
Castle
Casella Group
Cirrus
Cranfield Data Systems
Diagnostic Instruments
INVC

Sound level meters

Brüel & Kjær
Campbell
Casella Group
Castle
Cirrus
Larson Davis
Norsonic
Ono Sokki
Proscon
Quest Technologies
PC Werth

Testing

Anthony Best Dynamics
ISVR

6.3 CONSULTANCY SERVICES AND PUBLICITY

Consultants

Acoustic Associates
AEA Technology
WS Atkins
Automated Analysis

Brüel & Kjær
Burgess Industrial Group
Civil Engineering Dynamics
Hodgson & Hodgson Group
INVC
ISVR
Vipac

Publicity

Academic Press (Jnl of Sound & Vibration)
Institute of Acoustics
Multi-science Publishing Co. (Noise & Vibration Worldwide)
Society of Environmental Engineers (SEE)

Surveys

WS Atkins
Burgess Industrial Acoustics
Castle Group
Casella Group
INVC
Vipac Engineers & Scientists

Training

Brüel & Kjær
Diagnostic Instruments
Hodgson & Hodgson
Institute of Acoustics
INVC
ISVR

CHAPTER SEVEN

REFERENCE SECTION

7. Reference Section

7.1 GLOSSARY OF TERMS

[Terms in *italics* are further defined within the Glossary.]

Absorption - see *Sound absorption*
Absorption coefficient - see *Sound absorption coefficient*
Acoustics - The science of sound.
Acoustics of a room - Those factors which determine its character with respect to the quality of the received *sound*.
Acoustic calibrator - A device for producing an accurately known *sound pressure level*. Used for the calibration of *sound level meters*
Acoustic impedance of a surface, or acoustic source - The ratio of the *sound pressure* averaged over the surface, to the volume velocity through it. The volume velocity is the product of the surface area and *acoustic particle velocity*.
      Characteristic acoustic impedance (of a medium) - The ratio of *sound pressure* to *acoustic particle velocity* at a point in the medium during the transmission of a plane wave. It is the product of the speed of sound in the medium, and its density.
      Specific acoustic impedance (at a point in a sound field) - The complex ratio of *sound pressure* to the *acoustic particle velocity*.
Acoustic particle velocity - The velocity of a vibrating particle in an acoustic wave.
Active noise control - A noise control system which uses anti-phase signals from loudspeakers to reduce *noise* by destructive interference.
Airborne sound - *Sound* or *noise* radiated directly from a source, such as a loudspeaker, or machine, into the surrounding air (cf. *Structure-borne sound*)
Airborne sound insulation - The reduction or attenuation of *airborne sound* by a solid partition between source and receiver. This may be a building partition, e.g., a floor, wall, or ceiling, a screen or barrier, or an acoustic enclosure.

Aliasing - The introduction of false spectral lines into a *spectrum*, caused by having the maximum frequency of the signal greater than one-half the digital *sampling frequency*.

Ambient noise - The totally encompassing *noise* in a given situation at a given time, which is usually composed of *noise* from many sources, near and far (defined in BS 4142).

Analogue to digital converter (ADC) - A device which samples and digitises analogue signals, preparatory to *digital signal* processing. The continuously varying analogue signal is converted into a finite number of discrete steps or levels, and is represented as a series of numbers.

Anechoic - Without echo, i.e., without any *sound* reflections.

Anechoic room, or chamber - A room in which all the interior surfaces (walls, floor and ceiling) are lined with *sound absorbing materials* so that there are no reflections. It provides a standard environment for acoustic tests.

Anti-aliasing filter - A low pass filter inserted in an instrument, before the *ADC,* in order to prevent *aliasing*.

Attenuation - A general term used to indicate the reduction of *noise* or vibration, by whatever method or for whatever reason, and the amount, usually in *decibels*, by which it is reduced.

Attenuator - A device introduced into air or gas flow systems in order to reduce *noise*. Absorptive types contain *sound absorbing materials*. Reactive types are designed to "tune-out" noise at particular frequencies.

Audibility - The ability of a sound to be heard. The concept of audibility has been used as a criterion for setting limits to *noise* levels, particularly from amplified music. It is a subjective criterion, i.e., one which can only be determined by the ear of the listener, and not by measurement of *sound* levels. Also used as a criterion to determine the degree of privacy between rooms (e.g., offices).

Audibility threshold - The minimum *sound pressure level* which can just be heard, at a particular frequency, by people with normal hearing. Usually taken to be 20 μPa at 1000 Hz.

Audible range (frequencies) - Frequencies from 20 Hz to 20 kHz (approx.)

Audible range (sound pressures) - *Sound* pressures from 20 μPa to 100 Pa (approx.).

Audiogram - A chart or graph of *hearing level* against frequency.

Audiometer - An instrument which measures hearing sensitivity.

Audiometry - The measurement of hearing.

A-weighting - The frequency weighting dBA, defined in BS EN 60651:1994, which corresponds approximately to the human response to *sound*. It consists of a series of filters in a sound level meter. (cf. *C-weighting*).

Background noise level - The LA90 value of the *residual noise* defined in BS 4142.

Band pass filter - A filter which provides zero attenuation to all frequencies within a certain band, but which attenuates completely all other frequencies.

Band sound pressure level - The sound pressure level of the sound signal within a certain frequency band.

Bandwidth - The range of frequencies contained within a signal, or "passed" by a filter, or transmitted by a structure or device.

Bel - A non-dimensional unit, being the logarithm to base 10 of the ratio of two powers, such as sound intensity. See *Decibels.*

Broad band - A frequency band covering a wide range.

Centre frequency - The centre of a band of frequencies. In the case of *octave* or third-octaves it is the geometric mean of the upper and lower limiting frequencies of the band.

Characteristic acoustic impedance - see *Acoustic impedance.*

Continuous spectrum - A *sound* or vibration *spectrum* whose components are continuously distributed over the particular frequency range, as, for example, in the case of *random noise*, and in contrast to a line *spectrum*, from a *harmonic sound.*

Continuous equivalent noise level (LAeq) - See *Noise.*

Crest factor - The ratio of the peak of a signal to the *root mean square* value.

Criterion - The basis on which a noise or vibration is to be judged, e.g., damage to hearing, interference with speech, annoyance, etc.

C-weighting - The frequency weighting, defined in BS EN 60651:1994, which corresponds to the 100 *phon* contour and which is the closest to the linear or unweighted value. (cf. *A-weighting*)

Cycle - of a periodically varying quantity : the complete sequence of variations of the quantity which occurs during one period.

Cycle per second - Unit of frequency, also called Hertz (Hz).

Damping - The dissipation of vibration energy within a vibrating body, making it an inefficient acoustical radiator.

dBA - see *A-weighting.*

dBC - see *C-weighting.*

Decibel (dB) - A scale for comparing the ratios of two powers, or of quantities related to power, such as *sound intensity*. For example, if the difference in level between two powers, W_1 and W_2, is N dB, then $N = 10\log(W_1/W_2)$. The decibel scale may also be used to compare quantities, whose squared values may be related to powers, including *sound pressure*, vibration displacement, velocity or acceleration, voltage and *microphone* sensitivity. In these cases the difference in level between two signals, of magnitude S_1 and S_2 is given by $N = 20\log(S_2/S_1)$. The decibel scale may be used to measure absolute levels of quantities by specifying reference values which fix one point in the scale (0 dB) in absolute terms. (1 dB = 0,1 bel.)

Diffuse sound field - A sound field of statistically uniform energy density in which the directions of propagation of waves are random from point to point.

Digital signal - A signal which has a discrete number of values, which can be represented as a sequence of numbers. See also Analogue to Digital converter, and Digital to Analogue converter.

Digital to analogue converter (DAC) - An electronic device which converts *digital signals* into analogue signals.

DAT (Digital Audio Tape Recorder) - A tape recorder which includes an *ADC* (and a *DAC*) and which records analogue signals on tape in coded digital form.

Direct sound - *Sound* which arrives at the receiver having travelled directly from the source, without reflection.

Direct sound field - That part of the sound field produced by the source where the effects of reflections may be neglected.

Directivity factor - The ratio of the *sound intensity* at a given distance from a source, in a specified direction, to the average intensity over all directions, at the same distance (i.e., to the *sound intensity* at the same distance, when the sound source is non-directional).

Directivity index - The directivity factor (DF) of a source, expressed in decibels, i.e., 10log(DF).

Dynamic range - The range of magnitudes of a signal which, in a measuring system, or component of a system, can faithfully record, process or measure, from highest to lowest. Usually expressed in decibels.

Electret, or prepolarised microphone - A type of condenser microphone in which a prepolarised layer of electret polymer is used as a dielectric between the diaphragm and backing plate which form the condenser.

Electrostatic actuator - A device which fits over a *microphone*, close to the diaphragm, which is used for remote calibration.

Equal loudness contours - A standardised set of curves which show how the loudness of pure *tone* sounds vary with frequency at various *sound pressure levels*.

Equivalent continuous noise level (LAeq) - See *Noise*.

Fast time weighting - An averaging time used in *sound level meters*, and defined in BS EN 60651:1994.

Far field - That part of the sound field from a source where the *sound pressure* and *acoustic particle velocity* are substantially in phase, and the *sound intensity* is inversely proportional to the square of the distance from the source, i.e., the sound decays at 6 dB for a doubling of the distance from the source.

Fast Fourier Transform (FFT) - An algorithm, or calculation procedure for the rapid evaluation of Fourier transforms. An FFT analyser is a device which uses FFTs to convert digitised waveform signals into *frequency spectra*, and vice-versa.

Filter, Noise - A device which transmits noise signals within a certain band of *frequencies*, but attenuates all other *frequencies*. Filters may be electrical, mechanical or acoustical.

Flanking transmission - The transmission of *airborne sound* between two adjacent rooms by paths other than via the separating partition between the rooms, e.g., via floors, ceilings and flanking walls.

Fourier analysis/series/spectrum - Fourier's theorem shows that any periodic function may be broken down (or analysed) into a series of discrete harmonically related frequency components which may be represented as a line *spectrum*.

Fourier Transform - A mathematical process which transforms a non-periodic function of time into a continuous function of *frequency*, and vice-versa (in the case of the inverse transform).

Free field conditions - A situation in which the radiation from a *sound* source is completely unaffected by the presence of any reflecting boundaries. See *Anechoic*.

Frequency - of a sinusoidally varying quantity such as *sound pressure* or vibration displacement : the repetition rate of the cycle, i.e. the reciprocal of the period of the cycle, i.e. the number of cycles per second, measured in *Hertz* (Hz).

Frequency analysis - The separation and measurement of a signal into frequency bands

Frequency response of measurement system, or component of such a system, e.g., a *sound level meter* or a *microphone* : the variation in performance e.g., sensitivity with change of *frequency*.

Frequency spectrum - A graph resulting from a frequency analysis, showing the different levels of the signal in the various frequency bands.

Frequency weighting - An electronic filter built into a *sound level meter* according to BS EN 60651:1994. (cf. *A-* and *C-weighting*).

Harmonic signal - A signal which has a repetitive pattern.

Hearing level A - Measured threshold of hearing, expressed in decibels relative to a specified standard threshold for normal hearing.

 # Hearing loss - Any decrease of an individual's *hearing levels* above the specified standard of normal hearing.

Hertz (Hz) - The SI unit of frequency. The number of cycles per second.

High pass filter - A filter which transmits *frequency* components of a signal which are higher than a certain cut-off *frequency*, but which attenuates those below the cut-off.

Impedance - see *Acoustic impedance*.

Impedance matching - The use of a device to act as a buffer between a system, or component of a system with a high output impedance and one with a low input impedance.

Impulse - A transient signal of short duration. Impulsive noise is often described by words such as 'bang', 'thump', 'clatter'.

Infrasound - Acoustic waves with frequencies below the audible range, i.e., below about 20 Hz.

Insulation - see *Sound insulation.*

Interference - (1) The principle of interference governs how waves interact, with the combined wave disturbance being the algebraic sum of the individual wave disturbances, leading to the possibility of constructive and destructive interference. (2) Interference refers to the disturbing effect of unwanted signals, often electrical in nature.

Integrating sound level meter - A *sound level meter* which electrically integrates *sound pressure* signals to measure the equivalent continuous sound level, LAeq.

Intensity - see *Sound intensity.*

Level, L, sound pressure level, SPL - In general, the term *level* in this book implies the use of *decibels* related to the ratio of powers, or power related quantities such as *sound intensity*, or *sound pressure*.

LA - see *A- weighted sound pressure level.*

LAE - see *Sound exposure level, SEL.*

LAeq,T - see *Equivalent continuous sound level.*

LAmax - The maximum rms a-weighted sound pressure level occurring within a specified time period. The time weighting, Fast or Slow is usually specified.

LAN,T - Percentile level, i.e., the sound pressure level in dBA which is exceeded for N% of the time interval T , as for example in LA,10 and LA,90 .

Lpeak - see *Peak sound pressure level.*

LW - see *Sound power level.*

Level recorder - An instrument for registering and measuring the variation of signals, such as *sound pressures*, with time.

Linear A - (1) Measurement device A linear if its output is directly proportional to its input. In the case of a *microphone*, for example, this means that the sensitivity is constant and does not change with *sound pressure level*. (2) Linear SPL means unweighted.

Linearity - The degree to which a device is linear.

Loudness - The measure of the subjective impression of the magnitude, or strength of a *sound*, generally from 'quiet' to 'loud'. See *Phon* and *Sone.*

Loudness level - The loudness level of a *sound* is the *sound pressure level* of a standard pure *tone*, of specified *frequency*, which is equally as *loud*, according to the assessment of a panel of normal observers.

Low pass filter - A filter which transmits signals at *frequencies* below a certain cut-off frequency, and attenuates all higher *frequencies*.

Mel - A unit of *pitch*. The *pitch* of any *sound* judged by listeners to be n times that of a 1 mel *tone* is n mels. 1000 mels is the *pitch* of a 1 kHz *tone* at a sensation level of 40 *decibels*.
Microphone - A transducer which converts acoustic signals into electrical (voltage) signals.

Narrow band filter - A band pass filter with a small bandwidth, i.e., less than one-third octave.
Natural frequency - A *frequency* of free or natural vibration of a system.
Near field (of a sound source) - The region of space surrounding the source where *sound pressure* and *acoustic particle velocity* are not in phase, and the *sound pressure* varies with position in a complex way.
Noise - (1) Unwanted *sound*. (2) Unwanted signal (usually, electrical) in a measurement or instrumentation system. (In the context of this book, noise is used almost exclusively in the first sense.)

   Equivalent continuous noise level, LAeq - The steady noise level, (dBA), which, over the period of time under consideration contains the same amount of (A-weighted) sound energy as the time varying noise, over the same period of time.

   Noise criteria (NC) curves - A method of rating, or assessing internal (mainly office) noise, devised by Beranek, in the 1940s. It consists of a set of curves relating octave band *sound pressure level* to octave band centre frequencies. Each curve is given an NC number, which is numerically equal to its value at 1000 Hz. The NC value of a noise is obtained by plotting the octave band *spectrum* against the family of curves. In order to meet a particular NC specification the noise level must be either below, or equal to the SPL in each octave band.

   Noise index - A method of evaluating, or rating a noise, usually by assigning a single number to it, based on some combination of its physical characteristics (*sound pressure level, frequency* and duration) and other factors, such as time of day, or *tonal* or *impulsive* characteristics.

   Noise limit - A maximum or minimum value imposed on a noise index, e.g., for some legal purpose or to determine eligibility for some benefit.

   Noise rating (NR) curves - A method of rating noise which is similar to the NC system but intended to be applicable to a wider range of situations. The method is defined in ISO 1996 (parts 1-3); the NR system is particularly used for offices.
Noy - A unit of noisiness, related to the perceived noise level in PNdB by the formula

$$PNdB = 40 + 10\log^2(noy)$$

Nyquist frequency - The *frequency* which corresponds to half the sampling rate of digitised data, above which *aliasing* occurs.

Octave - The range between two *frequencies* whose ratio is 2 :1.

Output impedance - The *impedance* of a device measured at its output.

Overload - A situation in which a component or system is used beyond its range of linearity.

Overload indicator - A device which indicates when an instrument is likely to give incorrect reading because it is being *overloaded*.

Particle velocity - see *Acoustic particle velocity*.

Pascal - Unit of pressure, equal to 1 N.m^{-2} .

Pass band - A band of frequencies which are transmitted by a band-pass filter.

Peak - The maximum deviation of a signal from its mean value within a specified time interval.

Perceived noise level - The *sound pressure level* of a reference sound which is assessed by normal observers as being equally noisy. The reference sound consists of a band of *random noise* centred on 1 kHz.

Percentile level, LAN,T - The sound level, in dBA which is exceeded for N% of the time interval T, as for example in LA10 and LA90 .

Period (of a repetitive signal) - The time for one cycle.

Periodic signal - A signal which repeats itself exactly.

Phon - Unit of *loudness level* of a sound, being the *sound pressure level* of a 1 kHz pure *tone* judged by the average listener to be equally loud.

Pink noise - A random broadband signal which has equal power per percentage bandwidth, having a flat, i.e., horizontal, *frequency spectrum* when plotted on a logarithmic frequency scale. (cf. *White noise*)

Pitch - That attribute of auditory sensation in terms of which *sounds* may be ordered on a scale related primarily to *frequency*. The unit of pitch is the *mel*.

Plane wave - A wave in which the *wavefronts* are plane and parallel everywhere, and so the sound energy does not diverge with increasing distance from the source.

PNdB - The unit of perceived *noise* level.

Point source - An idealised concept of an acoustic source which radiates spherical waves.

Preamplifier - A circuit which acts as an electrical *impedance* matching device between a transducer with a high output *impedance* such as a *microphone* or accelerometer and the signal processing circuits of the *sound level* or vibration meter

Prepolarised microphone - see *Electret microphone*.

Progressive wave - A wave that travels outwards, from its source, and is not being reflected.

Psychoacoustics - The study of the relationship between the physical parameters of a *sound* and its human perception.

Pure tone - A *sound* for which the *waveform* is a sine wave, i.e., for which the *sound pressure* varies sinusoidally with time.

Random noise/ vibration/ signal - A *noise,* vibration or signal which has a random *waveform*, with no periodicity.

Real time - Measurement and analysis undertaken at the same time as a signal occurs.

Real time analyser - A device which is capable of analysing signals (usually in the *frequency* domain) in *real time*.

Reference value - Standardised values used as the basis for *decibel* scales of *sound pressure*, *sound intensity*, *sound power*, vibration acceleration, velocity and displacement.

Resonance - The situation in which the amplitude of forced vibration of a system reaches a maximum at a certain forcing frequency (called the *resonance frequency*).

Resonance frequency - The *frequency* at which the forced vibration amplitude in response to a force of constant amplitude is a maximum. For an undamped system the resonance frequency is the same as the *natural frequency* of the system, but for damped system the resonance frequency is slightly reduced.

Reverberant sound/reverberation - The persistence of *sound* in an enclosed space which results from repeated reflections at the boundaries.

Reverberant room - A standard acoustic test environment designed to produce diffuse *sound* conditions throughout the space.

Reverberant sound field - The region in an enclosed space in which the reverberant sound is the major contributor to the total *sound pressure level*.

Reverberation time - The time required for the steady *sound pressure level* in an enclosed space to decay by 60 dB, measured from the moment the *sound* source is switched off.

Room mode - a three dimensional *standing wave sound pressure* pattern, i.e. a mode shape, associated with one of the *natural frequencies* of a room.

Root mean square (rms) - The square root of the average of the squared values of a set of numbers. For a *sound* or vibration *waveform* the rms value over a given time interval is the square root of the average of the squared values of the *waveform* over that time period.

Sabin - Unit of sound absorption. 1 sabin is the amount of *absorption* equivalent to one square metre of perfect absorber.

Sabine's formula - The formula for predicting reverberation times of rooms.

Sampling frequency of a digitised signal - The number of samples per second.

Sampling interval - The time interval between samples.

Semi-anechoic - A room with *anechoic* walls and ceiling, but with a sound-reflecting floor.

Semi-reverberant - A room which is neither completely *anechoic* nor *reverberant*, but in between.

Sensitivity of a transducer - The ratio of output to input, e.g., for a *microphone* the sound sensitivity is usually measured in dB at 1 V.Pa^{-1}.

Signal to noise ratio (SNR) - A measure of the strength of a signal, indicating its magnitude relative to the background 'electrical' *noise* in the measurement system. Usually expressed in dB.

Silencer - A device for reducing *noise* in air and gas flow systems, either of the absorptive or reactive type. Also called attenuators, or mufflers.

Slow time weighting (S) - One of the standard averaging times for *sound level meter* displays, defined in BS EN 60651:1994.

Sone - The unit of *loudness*. The sone scale is devised to give numbers which are approximately proportional to *loudness*. It is related to the *phon* scale as follows: Phons = 40 + 10log2 Sones.

Sound - (1) Pressure fluctuations in a fluid medium within the (audible) range of amplitudes and frequencies which excite the sensation of hearing. (2) The sensation of hearing produced by such pressure fluctuations.

Sound absorbing material - Material designed and used to absorb sound by promoting frictional processes and thus generate heat. Examples are mineral fibre materials, or certain type of open-cell foam polymer materials.

Sound absorption - The process whereby sound energy is converted into heat leading to a reduction in sound pressure level. Also, the property possessed by materials, of absorbing sound energy.

Sound absorption coefficient - A measure of the effectiveness of materials as *sound absorbers*. It is the ratio if the sound energy absorbed (i.e., not reflected) by a surface to the total sound energy incident upon that surface. The value of the coefficient varies from 0 (for very poor absorbers and good reflectors) to 1 (for very good absorbers and poor reflectors).

Sound exposure level SEL or LAE - A measure of A-weighted sound energy, used to describe noise 'events' such as the passing of a train or aircraft. It is the A-weighted sound pressure level which, if occurring over a period of one second, would contain the same amount of a-weighted sound energy as the event.

Sound insulating materials - Materials designed and used as partitions in order to minimise the transmission of sound. The best materials are those which are dense and solid, such as wood, metal or brick, although lightweight panels can also be effective when in the form of double skin constructions.

Sound insulation - The reduction or attenuation of *airborne sound* by a solid partition between source and receiver. This may be a building partition (e.g., a floor, wall, or ceiling), a screen or barrier, or an acoustic enclosure.

Sound intensity - The *sound power* flowing per unit area, in a given direction, measured over an area perpendicular to the direction of flow. Sound intensity is measured in W.m^{-2}.

Sound intensity level (LI) - *Sound intensity* measured on a *decibel* scale : LI = 10 log(I/Io) , where Io is the reference value of *sound intensity*, I.

Sound level meter - An instrument for measuring *sound pressure levels*.

Sound power - The *sound* energy radiated per unit time by a *sound* source, measured in watts (W).

Sound power level (LW) - *Sound power* measured on a *decibel* scale : LW = 10log(W/Wo), where Wo is the reference value of sound power , 10-12 W.

Sound pressure - The fluctuations in air pressure, from the steady atmospheric pressure, created by *sound*, measured in pascals (Pa.).

Sound pressure level (SPL or Lp) - *Sound pressure* measured on a *decibel* scale: Lp = 20log(p/pO) where pO is the reference *sound pressure*, 20x10^{-6} Pa.

Sound reduction index (R) - A measure of the *airborne sound* insulating properties, in a particular frequency band, of a material in the form of a panel or partition, or of a building element such as a wall, window or floor. It is measured in decibels : R = 10log(1/t) ,where t is the *sound transmission coefficient*. Measured under laboratory conditions according to BS2750. Also known as transmission loss.

Sound transmission coefficient - The ratio of the *sound* energy transmitted by a partition, or across a boundary, to the *sound* energy incident upon the partition or the boundary.

Sound wave- A pressure wave in a fluid which transmits *sound* energy through the medium by virtue of the inertial, elastic and *damping* properties of the medium.

Specific acoustic impedance - see *Acoustic impedance*.

Spectrum, Frequency - A graph showing variation of *sound pressure level* (or other quantity) with frequency.

Spherical waves - An idealised model of how *sound* propagates in *free field conditions*, and used as the basis of certain *sound level* prediction methods.

Standing waves - A wave system characterised by a stationary pattern of amplitude distribution in space arising from the interference of progressive waves. Also called *stationary waves*.

Steady noise - *Noise* for which the fluctuations in time are small enough to permit measurement of average *sound pressure level* to be made satisfactorily without the need to measure LAeq using an *integrating sound level meter*. Defined in BS4142.

Stationary waves - see *Standing waves*.

Structure-borne sound - *Sound* which reaches the receiver after travelling from the source via a building, or machine structure. Structure-borne sound is transmitted very efficiently in structures, i.e., with very little attenuation, and is more difficult to predict than *airborne sound*.

Subjective - Depending upon the response of the individual.

Superposition, Principle of - The wave disturbances in a medium caused by different sources may be combined algebraically.

Threshold of hearing for a given listener - The lowest *sound pressure level* of a particular *sound* that, under specified measurement conditions, can be heard, assuming that the sound reaching the ears from other sources is negligible.

Threshold shift - The deviation, in *decibels*, of a measured hearing level from one previously established.

Timbre - The quality of a *sound* which is related to its *harmonic* structure.

Time constant - The time required for the value of a process or quantity, decaying exponentially with time, to reduce by a factor of 1/e, where e is the exponential number 2,7183..

Time weighting - One of the standard averaging times (**F, I, P, S**) used for the measurement of *rms sound pressure level* in *sound level meters*, specified in BS EN 60651:1994. Note: **F** - fast, **I** - impulse, **P** - peak, **S** - slow.

Tone - A sound which produces the sensation of *pitch*. See also *Pure tone*.

Transfer standard - A calibrated *noise* source designed to fit over a *microphone*.

Transient - A *noise* or vibration signal which is not continuous, but which decreases to and remains at zero.

Transmission coefficient - see *Sound transmission coefficient*.

Transmission loss - see *Sound reduction index*.

Ultrasound - *Acoustic* waves with *frequencies* which are too high to be heard by human ears.

Ultrasonics - The study of *ultrasound*.

Unweighted sound pressure level - A *sound pressure level* which has not been frequency weighted, sometimes known as the *'linear' sound pressure level* - symbol : Lp .

Waveform - A graph showing how a variable at one point in a wave (e.g., *sound pressure* or particle velocity), or vibration, varies with time.

Wavefront - the leading edge of a progressive wave, along which the vibration of the particles of the medium are in phase.

Wavelength - The minimum distance between two points in the medium transmitting a progressive wave which are in phase.

Weighting - see *A-weighting, C-weighting, Frequency-weighting, Time-weighting*.

7.2 TABLES

Tables are positioned within the text.
Please see the index for specific requirements.

7.3 STANDARDS

7.3.1 - British Standards Institution [BSI]

British standards can be accessed from their web site. Many BS standards are automatically incorporated from ISO standards, thus ISO 11200 becomes BS EN ISO 11200. However, certain BS standards, not originating from ISO standards, are mentioned within the text and their full titles are given below:

BS 2750

BS 4142:1997 - Method for rating industrial noise affecting mixed residential and industrial areas. [20 pp]
BS EN 60651:1994 - Specification for sound level meters [36pp]
BS EN 60804:1994 - Specification for integrating - Averaging sound level meter [48 pp]
BS EN 61252:1997 - Electroacoustics - Specifications for personal sound exposure meters. [34pp]

7.3.2 - International Standards Organisation [ISO]

There is a very wide range of ISO Standards on noise and acoustics. Many of the standards relate to the nuisance of noise rather than to the noise being an indication of the condition of the machinery. The following represents a small selection on a variety of topics which have some significance in machine monitoring and test arrangements.

ISO 31-7:1992 - Specifications for quantities, units and symbols - Part 7: Acoustics (2nd Edition) [1 pp]
ISO 362:1998 - Acoustics - Measurement of noise emitted by accelerating road vehicles - Engineering method [12 pp]
ISO 532:1975 - Acoustics - Method for calculating loudness level [18 pp]
ISO 1680:2000 - Acoustics - Test code for the measurement of airborne noise emitted by rotating machinery [28 pp]

ISO 1996:1982/7 - Acoustics - Description and measurement of environmental noise - Part 1:Basic quantities and procedures [5 pp], Part 2: Acquisition of data pertinent to land use [7 pp], Part 3: Application to noise limits [3 pp]

ISO 3743: 1995 - Acoustics - Determination of sound power level of noise sources - Engineering methods for small, movable sources in reverberant fields - Part 1: Comparison of hard-walled test rooms. [20 pp] 1995

    - Part 2: Methods for special reverberation test rooms [36 pp] 1997.

ISO 3744: 1995 - Acoustics - Determination of sound power level of noise sources using sound pressure - Engineering method in an essentially free field over a reflecting plane [40 pp].

ISO 3746:1996 - Acoustics - Determination of sound power level of noise source using sound pressure - Survey method using an enveloping measurement surface over a reflecting plane [38 pp]

ISO 3822-3: 1997 - Acoustics - Laboratory tests on noise emission from appliances and equipment used in water supply installations - Part 3: Mounting and operating conditions for in-line valves and appliances [8 pp].

ISO 4412-1: 1991 - Hydraulic fluid power - Test code for determination of airborne noise levels - Part 1: Pumps [20 pp].

ISO 4871: 1996 - Acoustics - Declaration and verification of noise emission values of machinery and equipment [14 pp].

ISO 6798:1995 - Reciprocating internal combustion engines - Measurement of emitted airborne noise - Engineering method and survey method [15 pp]

ISO 8297: 1994 - Acoustics - Determination of sound power levels of multisource industrial plants for evaluation of sound pressure levels in the environment - Engineering method [10 pp].

ISO 8579-1: 1993 - Acceptance code for gears - Part 1: Determination of airborne sound power levels emitted by gear units [16 pp].

ISO 9614: 1993 - Acoustics - Determination of sound power levels of noise sources using sound intensity

    - Part 1: Measurement at discrete points [19 pp]. 1993.

    - Part 2: Measurement by scanning [19 pp]. 1996.

ISO 9902: 1993 - Textile machinery acoustics - Determination of sound pressure levels and sound power levels emitted by textile machines - Engineering survey method [32 pp].

ISO 10844:1994 - Acoustics - Specification of test tracks for the purpose of measuring noise emitted by road vehicles [18 pp].

ISO 11200:1998 - Noise emitted by machinery and equipment - Guidelines for the use of basic standards for the determination of emission sound pressure levels at a work station and at other specified positions [26 pp].

ISO 11201:1995 - Noise emitted by machinery and equipment - Guidelines for the use of basic standards for the determination of emission sound pressure levels at a work station and at other specified positions - Engineering method in an essentially free field over a reflecting plane [24 pp].

ISO 11202:1996 - Noise emitted by machinery and equipment - Guidelines for the use of basic standards for the determination of emission sound pressure levels at a work station and at other specified positions - Survey method in situ [24 pp].

ISO 11203:1995 - Noise emitted by machinery and equipment - Guidelines for the use of basic standards for the determination of emission sound pressure levels at a work station and at other specified positions from the same power level [16 pp].

ISO 11204:1995 - Noise emitted by machinery and equipment - Guidelines for the use of basic standards for the determination of emission sound pressure levels at a work station and at other specified positions - Method requiring environmental corrections [26 pp].

ISO 11689:1997 - Acoustics - Procedure for the comparison of noise-emission data for machinery and equipment [26 pp].

ISO 11819-1:1997 - Acoustics - Measurement of the influence of road surfaces on traffic noise - Part 1: Statistical pass-by method [27 pp].

ISO 12001:1997 - Acoustics - Noise emitted by machinery and equipment - Rules for the drafting and presentation of a noise test code [30 pp].

7.4 REFERENCES

Abbott, P.G., Phillips, S.M. & Nelson, P.M. (2001) - Vehicle operation and road traffic noise reduction, *Noise in London Proc. of the Inst. of Acoustics*, May 2001, pp 53-60

Akishita, S., Li, Z.. & Kato, T. (1998) - Fehlerdiagnosesystem für Kraftfahrzeugmotoren, *Patent DE 198 22 908*, Nov. 11, 1998, Daifuku Co, Ltd., Osaka.

Dankwort, R. (1998) - Method and apparatus for non-intrusive, continuous noise monitoring, WO Patent 98/54544, December 1998.

Diagnostic Instruments (1996) - When it sounds right it cuts right, *Professional Engineering*, Vol., 1996, p 42.

Dings, P. (2001) - Railway noise reduction by controlling wheel and rail roughness, *Noise & Vibration Worldwide*, March 2001, Vol. 32, No.3, pp. 17-26.

Drives & Controls (1998) - Bearing noise analyser is completely automated, *Drives & Controls*, June 1998, p 6.

Eureka (1998) - Making all the right noises, *Eureka*, March 1998, p 26.

Harris, T. (1994) - Testing tyres to the limit , T.Harris, *Neural Edge*, No 6, Summer 1994.

Johnson, D.E. (1993) - *Use of noise for the detection of gear faults in rotating machinery*, Ph.D. Thesis, University of West of England, Bristol, March 1993, 350 pp.

Kokko, V. (2000) - Current situation and prospects of condition monitoring of electrical equipment during operation, *Proc. of Maintenance, condition monitoring and diagnostics - int. seminar*, Oct 2001, Oulu, Finland, Pohto Publications, pp 159-173.

Mucklow, P.A. (1970) - *The results of two tests applying acoustic diagnosis to gas turbine engines*, May 1970. Internal report Rolls-Royce Ltd, Derby.

N & V (2000) - The first tyre noise test was in 1847, *Noise & Vibration Worldwide*, Nov. 2000, p. 22.

Plant, R. (1998) - Sound advice, *Quality Engineering management*, February 1998, p 23.

Rasmussen, G. (1998) - Railway: wear and noise, *Proc. Institute of Acoustics*, Vol. 20, Part 1, 1998 pp 175-180.

Tomlin, G.M. (1984) - Quiet industry today - practical approaches to the control of noise at source. *Noise and Vibration Control Worldwide*, Feb. 1984, pp 80-83.

Vipac (1996) - "Bambino" wins coal research award, *Vipac News*, 1996.

Worley, S.A. (1990) - A practical diagnostic procedure leading to noise control by engineering means, *Engineering a quieter Europe*, Proc. I.Mech.E. London, 1990. C 406/016, pp 87-93.

Yang, S.J. & Ellison, A.J. (1985) - *Machinery noise measurement*, Oxford University Press. ISBN 019859 3333.

7.5 BIBLIOGRAPHY

7.5.1 Books

Barber, A. (1992) - *Handbook of noise and vibration control*, Elsevier Advanced Technology, 1992, ISBN 1-85617-079-9, 481 pp.

Serré, R. (1989) - *Elsevier's dictionary of noise and noise control*, Elsevier Science Publishers, 1993, ISBN 0-444-88073-9, 214 pp.

Turner, J.D. & Pretlove, A.J. (1991) - *Acoustics for engineers*, Macmillan Education Ltd, Basingstoke, UK, 1991, ISBN 0-333-52143-9, 192 pp.

7.5.2 Journals

Acoustics Bulletin (bi-monthly) - Institute of Acoustics

Applied Acoustics - Elsevier

Journal fo Low Frequency Noise and Vibration (quarterly) - Multi-science Publishing Co. Ltd.

Journal of Sound & Vibration - Academic Press Ltd

Journal of the Acoustical Society of America

Noise & Vibration Control Worldwide - Multi-scinece Publishing Co. Ltd.

Noise Management - Newsletter for professionals (monthly on web - www.noise-management.co.uk

INDEX SECTION

Editorial Index
Index of Authors

EDITORIAL INDEX

INDEX OF AUTHORS